Acknowledgements

I never really thought I would ever write and publish a book like this.

Now that I have, I need to reflect on how it happened and acknowledge and thank the people who have acted as influencers or providers of welcome assistance and encouragement along the way.

The motivation to get started came from Mary, whose interest and work in family genealogy led her to ask me to commit to writing a few of my memories of my parents and grandparents. I think she just meant a page or two, not two hundred and fifty!

Then along came Lisa who provided me with a little booklet called 'Grandpa – his stories, his words'. This set out a series of headings over vacant pages as prompts for a few handwritten sentences about my past. This was a good exercise for me at a time of recovery from prostate cancer surgery, and helped fill in a few weeks before getting out and about again. But it also stimulated me to think about what other headings I might want to write about – this story.

So, thanks to Mary and Lisa for acting as key influencers at the beginning.

The next step was to decide if this story might be of interest to others beyond our immediate family. Thanks here to my siblings Ken, Des, Shirley and Graham who provided positive feedback to a draft copy. And, for the strong encouragement I received from Lisa, Russell and Scott.

A very big thanks to Robert Blakely, a book club colleague here in Horsham who volunteered his time to very carefully proof-read and review my work, picking up on myriads of small errors, while also providing lengthy comments and suggestions for me to consider. Thanks, Rob, I greatly appreciate your assistance.

And finally, thanks to Ruth McLeod. Ruth is a self-published author herself ('From Our Mothers Arms', 2021), a fellow former student of Perenna State School, and childhood neighbour and friend from all those years ago. Thanks, Ruth, for volunteering to read and provide commentary on a final draft for me and for your subsequent encouragement and advice. An entirely appropriate person to write a forward for this book!

Disclaimer: This memoir is a recollection and interpretation of events as I saw them. If it is that I have misrepresented any event or persons within it, then please accept my unintentional error and fallibility!

Contents

Foreword - Ruth McLeod

Inspired by his mother, Margaret Dodds (nee Grimmer) writing about her early years, Ron Dodds compiled 'South of the Netting Fence: A Story of My 54 Years at Perenna'.

I could have been Ron's neighbour just south of the netting fence for 54 years. It didn't happen. When I was Ruth Evans growing up at Perenna, farming was a boy thing, and I found myself heading off to the big smoke for a different career. However, part of me never leaves and Ron's 39-chapter memoir is a treat.

Vignettes such as The Party Line, The Powerhouse, and The Nhill Show describe happenings that were part of my childhood too. In School Days, Ron pays tribute to our most influential teacher, Ron Attrill and recalls the highlights of the Annual Lake Hindmarsh School Picnic and Sports Day culminating in a rooster chase.

Ron's recollections are both detailed and precise. His descriptions in 'Living off the Land' and 'The Railway Yard' are familiar but also expand my knowledge. Many chapters like 'The Crop and Fallow Competition' introduce me to intricacies of farm life I knew little about.

Part 1 is Ron the farm boy learning farming, football and cricket from his grandfather and father, and not necessarily in that order with the tractor coming second to Saturday games and weekly practice sessions.

In Part 2, farming and community life are changing and Ron and his wife, Mary become the third generation to occupy the Dodds' family home.

Their tin-kettling might be one of the last, Guy Fawkes nights are petering out and new opportunities come along such as the Lake Hindmarsh Sailing Squadron. There is a notable loosening of the gender divide. The Saturday night dance in Netherby after the footy, once fragmented with the women inside the hall and men often hanging around at the front doors or going back and forth to the pub, rolls into a single event, the Cabaret Ball.

Honesty and a wry humour are features of innumerable anecdotes to delight the reader. Berly's horse droppings on their 'driveway' cricket pitch, Mary's first attempt lighting the wood-fuelled hot water service, the football chop night when the gas cylinder ran out before the keg was empty and air from a truck tyre substituted, are ones to look out for.

'South of the Netting Fence' is a treasury. Family history, farming developments and life in a 'community that was' are interwoven and upheld with good heartedness and an open love of country. It will be a conversation starter among his contemporaries. Dodds' descendants and younger readers have the opportunity to learn about the 'olden days'. Farmers, environmentalists, sports enthusiasts and all who enjoy a ripping yarn will marvel at Ron's commitment and knowledge.

Preamble

Given that a large part of my life – the first fifty-four years in fact – were spent growing up, living and working on the farm at Perenna, just south of the Vermin Proof Fence[1], it seems to me that this period of time is what really does make up my part of the Dodds family heritage, which may or may not be of interest to subsequent generations. It will be up to the reader to decide its value, but my thinking is if I don't bother to put my memories, or perhaps better to say my interpretation of events, into a readable form, then no one will ever be able to read it and decide it's junk or otherwise, so here goes!

The other reason to do so, in fact the catalyst, is Mary's keen interest in family genealogy – which has led her to trace and record the Dodds family line from their roots in Northern England (Newcastle-upon-Tyne), back to 1796.

While this in itself is interesting, my feeling is that the most important part of our heritage is really what life was like for our forebears. What was day-to-day life like for them? What did they do and think? What drove them to get in a flimsy sailing ship, and sail to the bottom of the world to start a new life there? The truth is, we don't really know, at least not in a 'firsthand' sense. The closest I have got to any understanding is to have sat at a dock side bar in Newcastle-Upon-Tyne enjoying a pint of English Ale and mused a while about the point of why they left there. I don't think it was religious persecution, so it must have been economic. Or was it pure adventure? Maybe they got sick of the cold!

The lack of any real recorded 'nuts and bolts' knowledge of our forebears is the key reason why we probably should all leave some sort of memoir. Then again, maybe it won't be necessary, because in this new data-filled-to-bursting world, all that will be needed will be to trace back through our Facebook page or our Twitter (X) posts. But my digital record is pretty patchy, and spans not much more than ten of my seventy plus years, so I will persist with this written account. I hope you like it.

To keep my thinking about my story of the farming way of life at Perenna in some kind of order, it seems to me best to set it out in two parts – by generation really. The fifties and sixties, the years of my youth and early farming days, and then the seventies and eighties, when in turn Mary and I raised our family, and the farming way of life of those times.

1 A wire-netting fence built in the late 1800s along the 36th parallel of latitude.

PART 1 – THE FIFTIES AND SIXTIES

Working the Fallow

After I left school in 1964 to return to the farm, I most enjoyed driving the tractor in the 'Hill' paddock on our farm at Perenna, twenty miles (Thirty-two kilometres) north of the then thriving town of Nhill.

It wasn't really much of a hill as hills go – maybe ten to fifteen metres above the general lie of the district. But enough to give the local landscape sufficient relief to provide interest – a view to a distant horizon, probably fifteen kilometres away, where the next similar low ridge added to the district's topography. These generally parallel ridgelines are relics of an ancient coastline, as the sea receded to the south-west in the geologic past; their existence makes the West Wimmera area, in my probably biased view, a more attractive landscape than the mostly flat plains that make up much of Australia's 'wheat sheep' belt.

That's why I liked the Perenna Hill – it broke the monotony of driving our Chamberlain '9G' tractor, towing a twelve-foot wide (sorry – make that about three and a half metres!) 'SunDuke' scarifier, (a cultivation implement) 'working the fallow' in preparation for sowing the next crop.

I could gaze across this landscape of square and rectangular paddocks, making up a patchwork of fallowed, stubble and pastured land, and could see just how busy and agriculturally industrious this district was. And as a counterpoint, if I gazed to the north, there was the low line of scrub covered sandhills – the southern edge of the Big Desert, or as my dad (Cliff Dodds) referred to it, the 90-mile desert, bordered by the 'Vermin Proof' Fence. Beyond the fence there was no farm or homestead for the next ninety miles. The Vermin Proof Fence or the 'Netting Fence' as we also knew it, was built in the late 1880s, for most of its route along the 36th Parallel of Latitude, between the Murray River at Tyntyder, with the purpose, as I understand it, of keeping native wildlife, or 'Vermin', away from the developing farming and grazing districts further south.

We will leave the Big Desert for later. To the southwest, rising in the distance above the horizon, was the concrete wheat silo, built back in the 1930s, at Netherby. And looking to the southeast, there was the Lorquon Silo, in this case matched with a more recent adjacent steel elevator and two steel silo bins. In the intervening foreground before these impressive landmarks, the districts roadways were laid out a mile apart in an east-west, north-south grid (magnetic north actually). Narrow lines of trees along these roads, and along fence lines between paddocks, with occasional larger blocks of 'timber', usually adjacent to the many farm homesteads, gave this landscape its particular appeal; that of a productive, peopled farming district.

As I rolled along with the Chamberlain 9G at full throttle, calling upon all of its sixty-five horsepower to drag the Scarifier up the slope to the top of the hill, there was much to see and think about. This little district was my home and workplace, and I have to say, I loved it. I must have, or surely, I would have taken heed of the Headmaster at Nhill High School, Mr Bishop, whose parting words to me upon my completion of my 'Matriculation' (Year 12) were 'there goes another young talent to be ploughed back into the soil'. His advice was to head off to the city, further my education, and become a teacher or public servant.

It may well have been that I would never have been sitting on that tractor were it not for the serious injury that Dad sustained in 1961 when I was just completing year 11 at Nhill High School. He very nearly lost his arm in an accident with the header, leaving him after many months of hospitalisation and recovery, with a significant disability, and greatly reduced ability to carry out much of the farm work he had previously undertaken. Were it not for that, it would most likely have been that he and older brother Ken, who already had left school and taken up a role on the farm, would have been an adequate labour force, and the options for me would have been different. Or maybe not – because a consistent memory of my childhood is that I was always told, or it was inferred, that I was a 'born farmer' – with a natural expectation that that was what I would become. I was never really inspired to think otherwise. I believe this was a common social expectation of the times – boys born into farming families, unless they showed a particular attribute or were deemed to be particularly clever or gifted, (Never Girls!) would follow their fathers and grandfathers before them and 'work the land'.

But for now, it's back to the unsprung, thinly padded bench seat on the 9G, and the view around me. Having turned the corner and heading west back down the hill – giving probably ten of those horsepower a temporary rest – our homestead, and the distant Netherby silo were in full view.

The silo was the only hint from this distance that gave a clue to Netherby's existence. There was indeed more to Netherby. For starters, there were four churches, with one 'parsonage' for the resident Lutheran minister. Balancing this, was the corner pub, plus a general store, a post office, a 'Mechanics Institute' , which had morphed into a Memorial Hall, a black smith's sprawling corrugated iron workshop, a motor garage, a stock and station agent, a railway station, a fire shed, a primary school, many houses, and of most importance, the local Football/ Cricket ground, with of course, a tennis and netball court. Netherby was quite a place of activity, in the midst of this farming community where you would find a farmhouse every mile or so, the surveyors of the late 1800s having laid out the district in square mile (640 acre) blocks. The population of the township itself was never large – but I imagine would have been in the hundreds.

That was the other reason why I liked working in the 'Hill' paddock - it provided a real sense of connection with this active little district and its various social structures. Every way I looked I would see other tractors busying their way across the paddocks, and Utes and cars driving up and down the roads. Probably just through the fence to the east, Evan Evans would be working away on his orange painted 'Chamberlain Crusader'. Another paddock or so to the north would be Sandy Morrison driving his red 'International A-554'. Further east I would catch occasional glimpses of Stan Deckert on his canary yellow 'Minneapolis Moline.' Back to the west – and there was Normie Smith on his blue 'Fordson Major'.

Normie's little tractor had a different character to the others – it didn't have a driver's cabin with louvre windows on three sides – his protection from the Northwest Wimmera Sun (and occasional shower of rain) was provided by a nice big umbrella bolted to the mudguard. South of Normie, Len Evans would be out on his red 'International AW-9', and then a paddock or so south of him, Ardie Nuske would be powering away on his Green 'Lanz Bulldog' with the unmistakable 'thump thump thump' of its single cylinder engine. No cabin or umbrella for Ardie – just a hat and coat! Across from Ardie, Merv Cramer would be out on his yellow 'Minneapolis Moline'. Wherever you looked, there was a farmer out driving his tractor.

Add up all of those machines – which we all saw as impressively powerful at the time, and you might reach three hundred horsepower. So, here's the core part of the story of my time of farming out north of Nhill. A single tractor used by today's farmers is likely a four to five hundred horsepower machine. How many tractors do you reckon you could spot now from the top of the Perenna Hill in the busy cropping season? If you're lucky, and be there on the right day, you might see one. Fast forward thirty years from my earlier time on the Chamberlain, I mused as I drove one of these larger multi-hundred horsepower machines that we had graded up to – how lonely it had become out there. There were no other neighbours to see close up. Maybe a pall of dust five or six kilometres away, where another similar machine was operating.

That active, busy little community, with a family at least on every square mile or 640-acre block, is gone, replaced by something very different. Mechanisation has led to massive social change, and depopulation of the area. Today, farming is still the key activity, and in fact crop yields are now at levels all those farmers of the 1950s would have seen as fanciful, such has been the advance of science and technology, not just with respect to mechanisation. Today's farms, in a good season, produce huge tonnages of grain compared with those earlier days, but the long-established declining terms of trade for agriculture has tightened profit margins. ('Terms of Trade' is the ratio between costs of farm inputs and prices received for farm products) This means that to be economically viable, farms have to be very large indeed, with the business preferably also diversified into other districts and other lines of income. This all means way less people living in the landscape.

The changed social scene could best be summed up by another feature I could view from the tractor seat up there on the Perenna Hill

That was the Perenna State School – number 4091, which ceased to exist as a place of learning in 1976. This was where I, and my siblings, Ken, Des, Shirley and Graham, all gained our primary education. As I drove up and down the paddock that day back in 1964, I watched the current crop of kids finding their way – variously by car, Ute and or bike, to and from their school day at the little timber schoolhouse, on a corner at the base of the slope, a mile from our homestead. Up until a decade or so earlier, the school had as its immediate neighbour on that corner, a simple little adobe or 'mud brick' building, which was the Perenna Methodist church, where we were all raised in the fear of God, with a church service every Sunday.

There were never big numbers of pupils at the school, or worshippers at the church – but the fact was they existed, because they served a then thriving little community.

What does that community look like now? Well, it actually doesn't exist anymore.

When Mary and I left Perenna in December 2000, (our children Lisa, Russell and Scott had left before us during the 1990s to complete their educations, and never came back) the population of the district was reduced to just one – Normie Smith, a single fellow, who still had his little blue Fordson Tractor. He was the only person left in the district that was once served by that little school and church, where all those farming men drove their individual little tractors, where all those families lived, where all those farmhouses stood, where all their machinery sheds, haystacks, barns, cow sheds and chook houses were important little elements of that industrious and prosperous farming area.

They are now all empty or gone. My story spans a time of great change in this rural landscape, and I hope gives some insight into a social structure that once existed but now has passed. My story is also one of I believe good fortune. For we were fortunate indeed that our childhood years coincided with what history shows to have been the key decades of boom times, socially and economically, in the wheat sheep belt – the 1950s and 1960s. And that the subsequent decades of the 1970s and 1980s, the second part of my story, were when Mary and I married, and we raised our family while seasons continued to largely favour us with a viable business, and a vibrant social life. But change was afoot, and really accelerated through the 1990's.

There is much that you inherit from your upbringing or formative years, and I celebrate the fact that I am a born optimist. For instance, I firmly believe that I will reduce my golf handicap to its lowest level ever in my 78th year, and that Collingwood will win the premiership again soon! (In fact, they just did!) I believe my inherited optimism is an ingrained leftover from the time. The visionaries who settled and developed the district were clearly driven by an optimistic hope of, and belief in, a grand future of prosperity. They built up towns and villages, which I am sure they saw as places of continued growth, social progress and development. And for a while, this dream was realised, peaking in those two decades of the 50s and 60s. They could not collectively have built and created the successful communities they did without optimism and a belief in the future. Critically, they were agents of change, willing to replace old ways of doing things with new. Willing indeed to relocate from the other side of the world to build a better future for themselves and their families. They embraced change in a way that unfortunately is now much less evident in many regional areas.

So how did the Dodds family get to Perenna in the first place?

I guess this is really the starting point, as stimulated by Mary's interest in genealogy, of this story. How did we ever get to be there in the first place, and why?

We got there because my grandfather, John Walter (Jack) Dodds, who was one of a family of eleven, raised in the back streets of Collingwood, Fitzroy and South Melbourne, left the city in 1904, and took up a job as a 'teamster' with the Dufty family on their property at 'Hope Glen'. Jack's forebears had arrived in Melbourne from Newcastle-upon-Tyne in Northern England in the late 1850s, chiefly plying their trade as boot and shoemakers and repairers.

The Hope Glen homestead was a large, rambling mud brick structure, built to reflect the Dufty's then dreams of lifestyle, the design even including a 'Ballroom'. It was located two miles east of the Netherby Township.

I don't know a lot about the Dufty family history, other than that they were a family who were amongst the earliest settlers in the district, taking up a holding of several thousand acres. It would be interesting to learn more of the Dufty's story that led them to take up their holdings out north of Nhill.

Many of the early 'squatters' who took up large holdings right across Victoria's Western Districts were of Scottish origin, who moved to take up big grazing runs soon after explorer Sir Thomas Mitchell returned to Sydney to report glowingly of the 'Australia Felix' he had traversed. The most desirable land was what we now know as the Victorian Volcanic Plain – a large sweep of land stretching from Melbourne across to the South Australian border, generally south of the Great Dividing Range. As Mitchell saw it, the volcanic plain was a mostly treeless grassy region. Being south of the divide, it enjoyed good rainfall, so presented itself as an ideal prospect for the grazing of cattle, and more particularly sheep.

So, it was a drawcard for the many Scottish families who took up land, and many of them did extremely well in this rich productive pastoral paradise. It's a whole other story, of wealth and privilege, and social separation based on that wealth[2]. The Western District was probably more than any other part of Australia a place of class division between the wealthy landholders and the working people whom

2 The Vanished Land – Disappearing Dynasties of Victoria's Western District, by Richard Zachariah

they depended on to maintain their lifestyle. Their time of great wealth came, and mostly went, for various reasons, to be replaced by others. The Squatter's legacy remains in the many magnificent bluestone homesteads that they built as a sign of the wealth that the period of Western District 'Squattocracy' afforded them. Many of these stately homes remain, nowadays most likely owned by families who have made their wealth elsewhere and pumped in the large amounts of cash needed to restore them to their former glory.

My latter work for Greening Australia afforded me a good opportunity to explore closely this part of Southwest Victoria, piquing my interest in further understanding its history.

It is an interesting part of my story, that while Jack Dodds came to be employed at Hope Glen, which has now disappeared back into the Mallee soil, that our Mum's father, Rupert Grimmer, was a worker at 'Narrapumalap' at Wickliffe on the Hopkins River, almost in the centre of the Western District., Rupert came from a working-class background similar to Jack Dodds. It's also interesting that unlike Hope Glen, Narrapumalap stands today as a restored Western district mansion, above the little town of Wickliffe, down in the Hopkins River valley, where Mum's childhood was spent.

The Dufty's must have arrived too late to have been favoured by the good fortune of taking up land in this fertile well-watered region and ended up instead on the edge of the Mallee country out north of Nhill. Much less rainfall, and no rich volcanic soil here! But land that was suited to farming and grazing nonetheless, but never to the extent that families were able to build huge mansions, with extensive gardens and outbuildings and sheds to house things such as Rolls Royce's and other trappings of wealth. Such significant wealth was not part of our background, or apparently the Dufty's. But Hope Glen must have been a good try.

I have scant recollection of the homestead that the Dufty's built at Hope Glen. It was quite large and did apparently include the ability to entertain guests and host special events, having its own 'Ballroom'. Not having the benefit of Western District Bluestone as a building material, they turned to what was in plentiful supply – dirt. Hope Glen, like many other district homesteads built at the time, was made of 'Mud Brick'.

Figure 1 'Hope Glen' Early 1970's, before it was demolished.

My main recollection of 'Hope Glen' was of the big old shearing shed. The Dufty holding steadily reduced in size, with blocks being sold off to newcomers to the district, such as Jack Dodds. But most of these new smaller farmers were croppers, (grain growers) and most did not build their own shearing sheds for their relatively small flocks. There was no shearing shed on our farm. So, the Hope Glen Shed remained the place where most of the district sheep were shorn. It was an 'eight stand' shed, powered by a big old engine connected to the shearing stands by a leather belt to the overhead spindles. It was a hive of activity for weeks at a time.

Jack Dodds

Figure 2 John Walter (Jack) Dodds. Date unknown

Why Jack left Melbourne in 1904 and relocated two hundred and forty miles to the Northwest Wimmera was I expect for economic reasons. Life would have been tough for a city working class family through the depression years of the 1890s, contrasted to the then surging development of the grain growing and wool industries with the encouragement of Government policies of closer settlement.

The Dufty's must have had quite a workforce. One of them was our Nanna, Elizabeth Evans who was employed as a maid. Her siblings that I knew about, or recall were her brother George Evans, and sisters Emma and Agnes. I don't know much else about Elizabeth's childhood, other than she married Jack at quite a young age. They lived for a while in a home (I imagine it was quite modest) several miles further north of Hope Glen just inside the Vermin Proof fence. I understand that both of their children (Agnes, or Aunty 'Bub', and Cliff, our father) were born while they lived there. Again, my information about day-to-day life for this part of our family history is scant – I do recall Dad speaking of it being mostly Mallee Scrub, with an abundance of Mallee Fowl. He claimed that an occasional Mallee Fowl egg made its way into their diet! It's an interesting contrast that some ninety years later, one of my proud achievements was the restoration of a two-hundred-hectare block of native vegetation as a biodiverse carbon sink, with the specific objective of recreating habitat and connections to other habitat for this now endangered species.

They moved from there to a property at Pijick, west of Rainbow, and then soon moved back to Perenna, or 'Pine Flat' where they purchased 1280 acres from the Dufty's, and apparently held clear title by 1930. It was there that they built the house that I spent my first 54 years in. I have no knowledge of how it was achieved, but compared to today's economic environment, it was perhaps remarkable that from a starting point as a farm labourer in 1904, Jack had the means to purchase a substantial farm holding, and build a modern new house on it, all in the space of 25 years. It seems to me that such advancement would not be possible today.

'Fafa' and 'Nanna' (Jack and Elizabeth) both lived on the farm at Perenna through my childhood years. When Dad (Cliff) met and married our mother Margaret

Grimmer, a second smaller bungalow style asbestos based 'cement sheet' house was built for Nanna and Fafa quite close to the house they built in 1929.

I recall Nanna as a shortish, plump woman, while Fafa was a tall, lean, quite handsome man. Nanna 'doted' on us kids (Ken, me, Des & Shirley, and later Graham), and so did Fafa. One of my clear memories of Fafa was that he would always go to Nhill on Fridays in his Ute and come home with a crate full of soft drink from 'Nhill Aerated Waters' for us. (This probably explains why we all had so many fillings in our teeth!) Nanna was a strict 'teetotaller', being I believe a member or supporter of the Christian Women's Temperance Union, members of whom apparently used to seek, with limited success, to have the local menfolk 'sign the pledge' to forgo the drink. I recall Nanna as a very 'Victorian' type lady who rejected alcohol and favoured a straight and narrow path of behaviour, whereas Fafa rather enjoyed a bit of fun, and a beer or stout or two, before he came home from Nhill! I'm pretty sure the odd bottle of 'Ballarat Bitter' or 'Abbots Lager' would have found their way into the back of the Ute along with the soft drink.

He loved sport and bought us cricket bats & footballs and lots of other things to play with. He would bowl tennis balls at us to hit for hours.

Jack and Elizabeth, (Fafa and Nanna) had two children only – Our Father Cliff, and Agnes, or as she was known to us, 'Auntie Bub'.

Dad left school quite young to work on the farm with Fafa. He used to proudly state that he got his 'Merit Certificate' before leaving school at Grade Eight. I understand his schooling was all at Perenna, which first took students in 1920. (State School 4091) I am sure he always wished he had a better educational opportunity, and a chance to do something other than farming. That said, I think he did enjoy farming when we were young. I believe Jack and Cliff were good farmers, growing wheat and oats, and clover pastures on which they ran sheep. The seasons were good through the 1950s and 60s, so we as a family did quite well – going off on holidays to the beach in Melbourne or Adelaide every summer after harvest, generally in a brand-new Holden car, traded in on the next model as soon as it was released. Dad was a committed and proud 'Holden' man.

Figure 3 Cliff Dodds. Date Unknown

Mum – Margaret Grimmer – also came from a relatively poor upbringing. Her father Rupert Grimmer worked, as did Jack Dodds, as a labourer for a wealthy landholding family, in this case in the favoured Western District (Wickliffe). The family was not able to afford more than a basic education for their family of five. But Mum was a bright student and won a scholarship to go to Alexandra Girls College in Hamilton, and then on to teachers' college in Melbourne. Along the way she overcame the difficulties of being a girl from a family of modest means in a setting with girls from mostly wealthy families. She used to recall how she only had one good dress when she first went to school in Hamilton hand-made by her mother, while the other girls had wardrobes full of outfits most likely from fashionable Collins Street outlets. Photos show Mum as a rather attractive young woman when she arrived at Nhill as a schoolteacher (at the small rural one- teacher school of 'Hamlyn'). She soon met Cliff, married, and I was one of five kids – that was the end of her teaching career!

Figure 4 Mum, with Des and Shirley

Mum had much to overcome in raising us out on the farm, not the least being isolated, and living in the same house at least for a time, and then in very close proximity, as our overbearing Nanna.

I wish I knew more about Mum's early years at Perenna, and indeed the Nhill district, other than having four kids in four years – with a bit of a break before Graham followed. Sadly, I had few really in-depth discussions with her about those early days, and how she really recalled it. It's clear that as was the accepted case at the time, the moment a woman married, any paid independent career such as her teaching role ceased. The war years had seen a loosening of this shibboleth for some, as women entered the workforce out of national necessity, but most reverted to a life of domestic duties as the men (those who weren't killed!) returned and social norms reverted. Equal opportunity was decades away!

She did relate some of her first experiences as the teacher at Hamlyn School, and how she lived as a boarder with the Barber family. Exactly how she met Cliff I don't know, but she did relate some stories of a social scene with other young local women, particularly about one party that involved several young trainee airmen who were then stationed at the Wartime Airforce training base at Nhill. Cliff wasn't one of them, he must have beaten the opposition!

She also recounted to me that in order to compare notes with a fellow young woman stationed as the teacher at Diapur, she had to ride her bike along sandy roads for twenty miles to get there.

Figure 5 My Collingwood Jumper!

She fought hard to ensure we all went to high school – against opposition from Nanna at the time, who apparently believed going off to School in Nhill was a ticket to certain moral decline, and that going to grade eight in primary school was all the education, as had Dad (Cliff) before us, one could possibly need. I, and we, owe so much for Mum's part in ensuring we had good educations, by the standards of the time in rural areas. I must also add that Dad too was a strong advocate for seeing his family well educated, in a way not afforded to him, despite the family pressure. Indeed, he took a strong interest and was Chairman of the Nhill High School council for some time during my school years.

Mum did have a difficult period in her life on the farm, suffering from bouts of depression, which was not well understood or managed at the time. I am sure that isolation was a contributing issue – she never learnt to drive a car, (many women did not drive then) and had no means to get about and do her own thing or meet others when she needed to. She was hospitalised for extensive periods and underwent 'shock treatment'. She largely overcame the illness in her later years.

Looking back, I now look at it as regrettable that Mum was in such poor health and lacking the care and support, she needed during my formative teenage years.

She outlived Dad by 24 years and became self-reliant and independent. She managed well on her own. She became a passionate and committed gardener, and also joined a local writers' group. She was an inspiration to me in both these passions, and I enjoyed her company in those latter years.

Dad's wish to do other things led him to many voluntary public life positions – local shire councillor, the high school council, farmer organisations, and his main passion – sporting clubs, football and cricket. He was never a particularly good sportsman himself- but he loved it and encouraged me to play football and cricket whatever way he could. His dream was that I might have played for Collingwood - definitely always no more than a dream!

Figure 6. The working bee that finished harvest for us that year.

He had a serious accident on the farm in 1962 - his arm was caught in the front of the harvester, which after several operations and many weeks in Hospital in Adelaide, left him with one severely disabled arm, and unable to do much physical work. He was lucky to have survived, apparently mainly due to Mum's quick thinking in applying a tourniquet to his upper arm to stem the bleeding before medical help could arrive.

His accident and subsequent disability was the main reason I left school in 1963 and came home to join Ken and work on the farm. A notable aside to his accident was how the whole district rallied to help, including a 'working bee' to complete the harvest for us. Just about every harvesting machine in the district arrived for an amazing one-day effort to finish the job. This was considered newsworthy as far away as Melbourne – with an article in the Melbourne Sun Newspaper – under the heading 'He sowed goodwill – and his friends reaped". It epitomised the well-known willingness of rural people to help their neighbours in need.

With Ken and I as the farm workforce, Dad became almost totally involved in his various public roles, particularly local government, which led in 1975 to him being awarded an 'MBE' – (Member of the Order of the British Empire). He was Lowan Shire President for several years, and also held the position of deputy chair on the Grain Elevators Board of Victoria, and on the executive of the Victorian Wheat and Woolgrowers Association.

He actually thrived in these new settings, enjoying I am sure his long dreamt of opportunity to experience life beyond the farm gate. His latter positions led him to spend considerable time in Melbourne, with either the Federal Hotel in Collins St., or the Commercial Travellers Association club in Flinders St as his 'boltholes'. He mostly travelled by train – the Overland Express, which ran one train each way every night between Melbourne and Adelaide. These trains used to 'cross over' at Nhill or Diapur in the middle of the night, around 2.00am. I spent many a night of interrupted bedtime waking up just after midnight to drive him into town to catch the train to Melbourne, or to be there to pick him up.

He sometimes used to drive new Holden Cars from Melbourne back to Nhill. He would pick them up at either Dandenong or Fisherman's Bend, the GMH plants, and 'run them in' for the local Holden dealership 'Bullen Bros.' with whom he had a close relationship. They were the glory days of Holden, and Bullen Bros

Figure 7 CT Dodds' first Holden - a 1948 model FX

rode the wave of prosperity in the Nhill district, selling and servicing dozens of new cars each year.

He never lived to see it, having passed away in 1988 due to Prostate Cancer at the relatively young age of seventy-three, but I am sure he would have been absolutely chuffed to have known that two of his grandsons, Russell and Scott, would have come to have worked for some time with Holden prior to its demise, in the Design and Engineering Departments respectively.

He certainly enjoyed the trappings that went with his life as an 'executive' in Melbourne. I was never much involved but can certainly recall a couple of occasions when I joined in for dinner at a city restaurant with some notable luminary of the Grains Industry where the wine was decanted, and the night grew long.

Cliff Dodds wasn't the only one who used farming organisations or local government involvement as a means of graduating to another path, often career changing, forging terms as state or federal politicians, or other roles. And sometimes an escape from the bush to a life on the tiles. But that's another story, and not really the subject of this memoir.

Cliff may have had his faults, being hardly a sensitive new age type with respect to gender roles, but he was a product of his time, and as my time goes by, I do respect his commitment to encouraging us and his clear understanding of the importance of education and learning. Considering his own rather poor educational opportunity, what he did manage to do in the public sphere was no small achievement.

And I also reflect that with respect to the running of the farm, he did step back and pass on management responsibility at an early stage, unlike most of his peers.

More about Jack Dodds (Fafa), and our way of life. Jack's time in farming straddled the change from horses to the era of mechanisation – and he used to struggle with modern (by 1950s standards) machinery, cars, and tractors. Horses didn't disappear from his interests, as he remained a very keen follower of 'The Trots' – harness racing. Our northern Neighbour Eric Berlund was also a very keen trots man, and in fact, owned trotting horses which he used to train on a track in his front paddock. Jack would head off with Eric ('Berly') to the trots somewhere whenever they had the chance. In fact, it was his love of Trots that brought about his early demise in 1961, when he was killed in a head-on crash on the Western Highway at Rockbank. He and his good friend Bob Lineker had been to the Trots at Moonee Valley, but he never made it home.

For me, the most memorable aspect of Jack's friendship with 'Berly' was its interference with our cricket on the driveway beside the house at Perenna. Our Cricket Pitch was located just outside the garden gate near the kitchen, on a fairly flat area of loamy soil, nicely packed down by all the farm vehicles as they pulled up at the gate. Ken and I were the keen cricketers of the family, and we spent hours out there. The 32-volt power shed was conveniently located to become a good backstop in lieu of a wicket keeper, so apart from the gentle sideways slope to the east, causing a natural leg break, it was just about ideal. It might have been a bit close to the house with its 'cement sheet' walls, as evidenced by the covered-up holes in the wall where a few well-hit pull-shots through midwicket cleared the garden fence, but really just about ideal for a kids' cricket pitch.

The biggest problem we had though was when Berly arrived, almost daily, with his horse and jinker, and pulled up right in the middle of the pitch. Jack, and Cliff were always up for a yarn about pretty much everything, so with Berly sitting on the Jinker, and them leaning on the side, or patting the horse, and hour would pass by really quickly! The end result being several heaps of horse manure, pretty much right on a good length. We had to shovel this out of the way when Berly left, before the resumption of play, and the pitch always was a bit dodgy until the sun had dried it out again!

The Trots weren't Jack's only sporting interest – in fact it was probably (only) ranked second. Nothing outranked the fortunes of the Collingwood Football Club! He was a passionate supporter of the Magpies, as was Cliff, who in turn passed it on to us as well. In fact, if there is one thing that tracing the genealogy of the Dodds family does, is it links us to the back streets of working-class Collingwood in the 1890s, right at the birth of what is now the AFL. Jack and his siblings were raised there, and the success of 'their' football team was the common bond, as it was for thousands of others who used to crowd into Victoria Park (Collingwood's home ground) for every home game.

Jack had three brothers who never left Melbourne – our uncles Jim, Harry and Bert. They actually all became tram drivers, and as such were all staunch union men. I can well remember how their common dislike for the then Liberal Party Prime Minister Sir Robert Menzies almost always got raised when we visited them, (Uncle Bert in particular!) while Jack and Cliff's Country Party leanings were on the opposite side. So, they always changed the subject as quickly as they could to football to avoid ruining the night. Unlike politics, barracking for Collingwood was a subject of no disagreement!

I had far less contact with two other brothers of Jack – our Uncles Herb and (Uncle) Art, and never recall ever meeting his other siblings. As it turned out, my first and only meeting with Herb and Art was at the time of Fafa's untimely death and funeral, when they both came and spent some time with us. Herb travelled across from Western Australia. Herb had apparently moved up from Melbourne at around the same time as Jack to also work for the Dufty's, and for a while farmed a Mallee block north of us – out right near the Vermin Proof Fence, before setting out to find a better opportunity in the West Australian Wheat Belt, at Wongan Hills. I knew a bit about Uncle Herb, as Nana and Fafa had earlier taken a trip via the then 'Transcontinental Railway' to visit him and his family over in the west. I well recall meeting Uncle Herb, remembering most of all how well dressed he was, with suit and tie, and a neat felt hat, even for the few days he spent with us on the farm after the funeral.

But who was this other gentleman– Uncle Art (short for Arthur) as Dad introduced him to us? I knew nothing whatever about him, or that we even had such an uncle. And what a character he was! He was a few years younger than Herb and Jack, and was a fit and active, slim sort of bloke. I remember those few days well all these years on with Uncle Art, as he kicked the footy with us, regaled us with stories of his prowess at cricket, hitting sixes and bowling outswingers as an all-rounder for the Dumbalk Cricket Club in Gippsland, and spent time just hanging around with us. He was full of fun and a very interesting bloke as far as I was concerned! But why didn't I know about him beforehand? Well, it turned out that he had been banished from our family circle by Nanna, never to be spoken about, because, shock and horror, he had never married but was 'living in sin' with a woman on a dairy farm at Meeniyan in Gippsland! It took the death of his brother for him to be revealed to us. What a shame!

Returning to the football, family legend has it that Uncle Jim took up the same seat in the 'Raine' Stand at Victoria Park (Collingwood's home ground) every Saturday without fail and did so for years on end. He never missed a game. But, in a tradition still followed by some supporters today, couldn't stand the tension of the final quarter, with the risk of witnessing a loss, so he apparently always left and walked home before the final siren. Knowing this, our Auntie Anne, who would have the radio tuned to the footy, would have a black and white scarf hanging on the front gate if they won, but not if they lost.

Jack and Cliff also took every opportunity they could to get to Melbourne to see the Mighty Magpies play. This, according to Dad, (I believed his story!) extended to the pair of them hitching a ride down the Western Highway to Melbourne on the back of a truck load of wool to get to see one of the depression years grand finals. The depression years coincided with Collingwood's glory days – six premierships in ten years! It was a positive success to feel part of and share with your fellow supporters during grim economic times.

If they couldn't get to see Collingwood, there was always the local 'Lowan Star' League to follow – every little district centre could easily field a team every Saturday – and a netball team for the girls and women as well. Netherby, and Lorquon each had an active football club, as did Yanac, Boyeo, Winiam, Gerang, Kiata and Antwerp. Pretty much the whole district would turn out of a Saturday to see the local team play their visitors, almost always followed up by a Dance at the local Hall. They were big days of local socialising.

Unless, used as we were to Big Time League football, -VFL, you chose to go off and support your local Wimmera League side, in our case Nhill – a higher-grade competition where occasionally a promising young talent would be recruited from one of the Lowan Star teams. The Nhill team was hard to break into though, as with a population of over two thousand in the town itself, at a time when practically every young boy was expected to play footy, there was a decent talent pool to draw on.

And if for whatever strange reason you didn't go to any footy match on a Saturday, then it would be the time to tune in the radio and listen to Collingwood play. The advent of portable radios was a most important breakthrough in that you could listen to the footy away from the house, around the farm! Until you broke it – like the Saturday afternoon that Fafa decided he would sow some oats in the cow paddock near the house. This meant that I was a close spectator of the activity and remember how impressed I was by his initiative of having the Phillips Portable Radio (or 'Wireless' as we referred to radio's back then) sitting on the tractor mudguard, tuned into station 3LK where they broadcast the footy from Melbourne. Intent on the broadcast as he drove around the paddock, he missed seeing a bumpy patch, which sent the wireless off the mudguard down onto the ground, through the cultivating tynes of the combine, under the covering harrows and out behind the machine before he could stop. I recall running over just as he picked it up and shook the dirt off it, then turning the on/off knob to try and get it working again, because apparently Collingwood had been on the verge of kicking another goal! All to no avail. A new portable radio was on the shopping list for the next trip to Nhill – not to be sat on the tractor mudguard again. As a postscript, it was to be another twenty years before we had such a thing as a tractor with a driver's cabin in which we installed a real car radio!

The Bedford Truck

With respect to Jack's difficulty in traversing the change from the horse and buggy era to the world of new technology, I can remember several events that actually were quite funny: one in particular stands out.

It was harvest time, and one of the proudest new purchases for the farm was the brand new 1953 Five Ton Bedford Truck. It made quite a statement, painted a nice dark green, with black mudguards, and neat yellow and white pinstripes highlighting the doors and metalwork edges. The pin striping even carried over to the steel 'Sides' that converted the open tray to a modern bulk bin, marking the beginning of the age of 'Bulk Handling' on the farm. The driver's side door had a very professional sign writing job in big bold letters: -

'FW & CT Dodds,

Primary Producers,

Lorquon West'

Why they chose to nominate Lorquon West as the location of the farm and emblazon that on the door has remained a mystery to me, because there was really no such place. There were named locations all over the area, almost all with their own State School as an absolute minimum to denote a local community. There was Lorquon, Netherby, Baker, Hamlyn, Netherby South, Ni-Ni, South Lake Hindmarsh, North Lake Hindmarsh, and of course Perenna, just a mile from our Farmhouse. But no Lorquon West! Perhaps the name was chosen for the truck door because it carried more gravitas than 'Perenna', denoting a connection to a centre of much more importance and commercial activity, showing that the truck belonged to a place of substance, as Lorquon was at the time. If there is a location of Lorquon West, it would have been five miles south of the farm, not one mile like Perenna.

Anyway, we were all very proud of that truck, for which a brand-new shed was built on the end of the existing 'Machinery Shed', complete with two big white painted wooden doors mounted on big hinges, so the truck could be safely kept in out of the weather.

The builder of this extension to the shed, and the shed itself in the first place, and Nanna and Fafas' new house was Jim Curkpatrick, a carpenter from Nhill. All of this building activity meant that Jim, and his offsider whose name I cannot remember spent quite some time on the farm. I recall as a kid watching and following this work. I am sure that was my first introduction to the art of swearing, which of course was rather new and interesting, given Nanna's strict Victorian ways which she tried hard to instil into us! The sound of the hammer banging in nails, the handsaw cutting the timber to length (no power tools then) were strongly punctuated by lots of new words previously unknown to me!

But back to the story about the Bedford truck. The move to bulk handling at the time went only so far, because the harvester did not have such a thing as an auger to unload the grain, which was instead emptied out into Jute Bags, which were left with their tops left open in little heaps out across the stubble as harvest progressed.

Just under the side of the truck, beneath the tray, a 'G Well' hydraulic bag loader was installed. To this unit, you were able to attach with a couple of big metal bars the bag loader unit itself, which was a rather mean looking device with long metal arms, and a 'bucket' mounted on an angle on the arms, into which you hefted the bag of wheat, having dragged it's deadweight across the stubble, where four very sharp and dangerous looking pins pierced the bottom of the bag, to secure it. You then gave the bag a good shove with your knee, which in turn depressed a lever which actuated the hydraulic ram, and with the Bedford set at a fast idle, the loader would swing up with a jerk through about a 120-degree arc, at the top of which the bag was inverted, and emptied out into the tray of the bulk bin. As the bag was now empty, it released the pressure on the plate that you had actuated with your knee, releasing the hydraulic pressure, upon which the loader would swing back down again, crashing onto the ground with a bang, sending dust and stubble flying, ready for you to pick the empty bag off those sharp prongs, and drag up another full one to repeat the action. After doing this for the next eighty or so one hundred- and eighty-pound (seventy kg.) bags, you had the five-ton Bedford full of wheat. What an amazing labour-saving device! To say nothing of the occupational health and safety issues involved!

This leads me to the next diversion on this story of Fafa and the truck; how lucky were us kids with the freedom and exposure to risk we had. I probably mean how lucky I am to be still here to recount this real-life experience as a kid. We (I) loved the grain harvest! Particularly this loading of the truck with the G Well loader, because we could help, and have great fun while we did. No, it wasn't child labour, it was just fun!

There were two jobs to be done. One, we could stay down on the ground, and as soon, or even before, the loader crashed back to the ground, we could run in and grab the empty bag, pulling if off those needle sharp prongs, and stacking it neatly to be taken back to the header to be refilled again, allowing Dad or Fafa to just concentrate on the labour of dragging up a full bag, lifting it onto the prongs and into the bucket, and shoving their knee to send it on its way.

The other job was to clamber up into the bin on the truck, and scrape and push the pile of wheat, as it came in bag after bag, into the corners of the bin, to enable more to be poured into a load. The wheat was warm and smooth, and moved easily, and smelt beautiful. It wasn't hard work; it was like playing in a sandpit on steroids! Wheat pouring all over your head if you accidentally (on purpose) got in under the G Well as it reached the top, and then extracting yourself from the pile. But the best part of all was when the truck was half full.

This was because the mounting point of the loader was towards the front of the tray, which meant that the pile of wheat was mainly in the front, with the back of the bin still almost empty. (A pile of wheat rests at an angle of about thirty degrees). The fix was the fun part, because one of the other new technological aspects of the Bedford Truck was its 'G Well' underbody hydraulic hoist. In other words, it was a tip truck. So, to move most of the wheat down to the back of the bin, Dad or Fafa would leave the loader on the ground, take a much needed swig on the water bag, and climb into the driver's seat to engage the hoist, which would then tip the tray up, with us kids in the bin yelling out "More, more!", until the wheat would begin to slide towards the back when we would yell "Stop", while beginning to slide down along with it. We would be half covered in wheat when we stopped – the ultimate kids slide experience! If we didn't yell 'stop' quick enough, or if the response down in the cabin was too slow, too much would slide back, and overflow over the end. Not a good outcome! For most part, this was avoided with the hard-won wheat staying in the truck. With the wheat moved to the back, the hoist was lowered back down, and more bags were loaded while we scraped the wheat to the sides to make a nice smooth (and overloaded!) bin full.

It goes without saying of course that this was a dangerous thing for kids to be allowed to do, despite our enjoyment of it. Local children's lives were lost in similar situations when the grain pile they were allowed to play in while it was being loaded or unloaded shifted unexpectedly. Crushing and suffocation is a real risk. Suffice to say that the next generation of kids raised on the farm – ours – were never allowed such licence.

Now back to my original story about Fafa and the truck. It related to one summer evening at harvest time. The important background to this particular event is to explain a little bit about the layout of the farmyard. The two houses and sheds were all set back a bit from the road along the front. There was a gateway in the fence along the road as a main entrance. Along this fence line, strung on wooden poles was the telephone line, which passed over the top of the gateway, probably four metres or so above the ground. This was an adequate height for most of vehicles and tractors that used it, but unfortunately not quite high enough if the 'G Well' loader was left attached to the truck in its upright position. Further in the yard along the track in, the thirty-two-volt power line was also strung at about the same height from the Powerhouse (More about that later) across to Nanna and Fafa's new house.

Now there were a couple of reasons why you would leave the G Well Loader attached to the truck in its upright position. One is you were too buggered from dragging eighty heavy bags of wheat across loose dirt and stubble to shove on the loader, and just wanted to knock off. You didn't really want to pull out those two heavy metal bars while holding the loader, and then pulling it aside. It was easier to loop the piece of chain provided around the loader and clip it on –

remembering that the truck was now about a foot wider, and five feet higher! Or you needed to move operations to another paddock the next day, so the loader needed to come along.

Anyway, on this particular night, Fafa chose the first option. I can remember it well, as I was playing out in the garden by our house. The truck came noisily revving flat out in low gear up the road, as Fafa wasn't all that great on gear changes, which he avoided as much as possible. He slowed at the gate as he turned in, crunched into lower gear, and revved up again. Two things then happened – the loader hit the gate post, pushing it sideways, and the phone line was caught and stretched as he continued in a chain[3] or so before the wire broke with a twang above the noise of the engine, and he came to a halt.

The racket brought Nanna out onto her front porch, enquiring above the swearing and muttering as to what had happened. I think the answer was something like "Never mind Woman – Ill bloody well fix it up tomorrow", at which time he climbed back in the cabin, crunched it into gear again, revved it up, and proceeded forward until the loader collected the thirty-two-volt power line! This time I can recall more twanging of wire, and the pulling of the electrical insulators from the wall of our house and the top of the powerhouse. There was no phone, or electricity, on the farm that night!

---

3 A 'chain' was an imperial unit of measure, made up of twenty-two 'yards' (approx twenty metres). For context, a cricket pitch is one chain or twenty-two yards long.

Haystacks, Chaff and Crushed Oats

Jack Dodds might not have been great with newfangled machinery, but he had some real skills – such as building haystacks out of sheaved hay with a pitchfork.

Haymaking just preceded the grain harvest on the farm, generally taking place in October, depending on the season. We made two types of hay, sheaved and baled. The first was 'sheaved' hay which was made from an oat crop, grown for the purpose. The oats grew tall, with nice thick grassy stems, ideal for chaff-making. Oaten chaff was a key part of the diet of the working draught horse teams that were so important to the early days of farming in the area. They were almost totally superseded by the development of tractors by the Second World War. I have just one fleeting childhood memory of a team of horses, being driven by Fafa along the track up Perenna Hill.

The sheaved hay was produced by a 'Reaper Binder'. This machine was towed through the crop by a tractor (used to be horses), its mechanism driven by 'ground drive', which entailed a big metal wheel, to which a continuous chain was mounted, providing the power to turn a series of shafts. It cut off the crop just above ground level with a reciprocating serrated knife. The cut crop would then fall onto a continuous canvas elevator belt, which lifted it up and separated it into bundles, with a length of binder twine wrapped round it and a knot tied to form the 'Sheaf' of hay, which was then dropped onto the ground behind the machine. To me, it was an amazing machine, of which I was never able to even come remotely close to building a replica with my 'Meccano' set, that looked anything like the real thing. (I was also an abject failure in my attempts to produce a model of a "Hannaford Pickler' machine out of the huge array of Meccano pieces I had accumulated over several Christmases! I did manage to build something that might almost be recognised as a model of a 'Header').

How the binder managed to wrap that twine and tie a knot still eludes me, despite having spent some time studying one in a machinery Museum recently. The mechanisation of agriculture involved the manufacture of some pretty cleverly designed machines, some dating back into the 1800s, of which the reaper binder was one example.

Back to Jack's haystack skills. This involved first going down to the Pink Lake near Dimboola, a highly saline shallow lake which dried in spring and summer to leave a crust of salt, which was valued for many purposes. He would come home in his Ute with several bags of salt. Then it was out into the paddock with the Bedford truck, where the sheaves of oaten hay had been allowed to dry or cure in little heaps or 'stooks' for a couple of weeks after the binder went through. The sheaves would be loaded by pitchfork (Fafa & Dad), and then back to the haystack site near the cow shed. I don't really know exactly the dimensions of the base of the haystack, but it was probably about 10-15 metres long, by 4-5 metres wide. It had a low flat sheet metal fence around it, for the purpose of keeping mice out. With the truck parked beside the stack site, Dad tossed the sheaves down (and subsequently up) to Fafa, and he would steadily build the haystack with his pitchfork, placing and overlapping them neatly and consistently, thatching them all together so that none slipped or fell out of place, layer by layer, truck load after truck load. As he added each layer, he would spread handfuls of the Pink Lake salt, claiming that would 'sweeten the hay' for the animals. The stack sloped gently outwards as it got higher, up to about 4 metres, before it sloped back to a peak at the top, so that water would shed off the sides when it rained. When completed it looked a bit like a 'high tin' loaf of bread. One of my great regrets is not having a photo of one of the haystacks that Jack built! Perhaps the closest I can suggest is to view a Claude Monet work – his famous series of paintings were of stacks of hay of similar appearance.

What happened to the haystacks, and the use to which the stored hay was put was another interesting part of our life and experience as kids. The sheaved hay system was largely a left over from the horse era – the hay being cut into chaff which was a major part of the fodder consumed by the working horse teams – augmented by crushed oats. It was retained, in a scaled back manner, for the 'house cows' that provided us with the milk, cream and butter for our diet.

Both of these products (Chaff and Crushed Oats) were produced on the farm, from the sheaved hay in the haystack, and from bags of oats which were stored in the barn after harvest. The day the chaff was cut, and the oats crushed, was always a great spectator sport if we happened to be not at school at the time.

It meant a couple of rather wondrous pieces of machinery, located beside the corrugated Iron 'Chaff House' would be put into action. Machines like these are now of course museum pieces, but well worth a visit to an appropriate museum, if for no other reason than to be amazed by the dearth of operator safety considerations!

The Chaff Cutter had at its centre a big heavy metal wheel, to which were attached three big scythe like razor sharp blades, which rotated at speed. To feed the hay into these blades, whereby the straw would be cut into millions of tiny pieces about a centimetre long, was a trough which had two rather evil looking serrated metal rollers. As they rotated, they grabbed the straw, as the operator loosened the sheaf apart and fed it in by hand, to be forced in and swiftly cut by the big rotating wheel and blades. The resultant chaff then dropped down to the base of a long enclosed wooden elevator, which had a continuous leather belt with wooden paddles on it. This elevator was set at about a 45-degree angle and ran up through a hole in the wall of the chaff house, where the chaff was dropped into a heap which grew as the operation continued.

The oat crusher was positioned quite close to the chaff cutter. It's working parts were just two flat metal rollers, about half a metre in diameter, set to run very close to each other (about 1mm). A conical bin sat on top, allowing the bags of oats to be manually lifted up and poured in, and then fed through an adjustable metal slide down between the fast-rotating rollers, which crushed the oats, almost exactly like the rolled oats you buy in the supermarket today. The resultant crushed oats were then scooped up and placed back in the bags from which they came.

I can remember as a kid being mesmerised by watching these machines in action and listening to the noise it all made!

The motive power for this whole process was provided by the farm tractor – a Massy-Harris 55K. The early tractors didn't have what is now known as a 'power take off' or PTO shaft, but rather a round metal belt pulley about 30 cm in diameter. So, the first part of the exercise, was for the tractor to be driven down to the chaff house, and lined up in front of the chaff cutter, so that the tractors belt pulley was as near as possible exactly aligned the metal pulley on the chaff cutter.

Dad would be on the tractor inching it forward, with Fafa calling "Over to the Left!" "More!" "Further over" "No, that's no good! Back-back and come in again!" When it was about right, a big leather belt, probably about five metres long and about twenty centimetres wide, was then unrolled, and fitted between the pulleys on the tractor and the chaff cutter.

The next step would to be to tension the belt. This too, made for good spectator sport. It was done by Dad reversing the tractor until its tightness was deemed to be correct by Fafa, who then shoved a big wooden block under the tractor's front wheel to keep the belt tight. (The Massey Harris 55K tractor wasn't well equipped in the brakes department) A very careful but fraught operation! Dad would ride the clutch, with Fafa directing him to "Back" "Back" "A bit more" If Dad drove the tractor too far back, it would begin to tilt the chaffcutter off its rear foundations (a big block of wood), which would in turn move the elevator out of its proper place.(and start to shake the wall of the chaff house) That would cause some consternation! But he mostly got it right and engaged the Belt Pulley to see if it all ran OK. This was where the alignment issue was put to the test, because if it was out of line to any degree the belt would run off, and the whole procedure would have to be repeated with the tractor moved over a bit.

I forgot to mention too, that in installing the belt, it had to be 'crossed', because for some reason, if the belt was just put on straight, the chaff cutter would go backwards. I have a good recall of this, as once in operation, every time the belt did one rotation, the metal joining piece would go 'Thwack' as it wiped past the other side of the belt, crossed as it was.

With the chaffcutter all set up – including the handy presence of a big oil can with a spout on it, ready to be poured every few minutes into the flange bearings on the whirring drive shafts before they overheated, the next step was to drive the truck over to the haystack and pitchfork enough sheaves on to make a decent heap of chaff. This part of the operation also provided plenty of entertainment for us kids. This was largely due to the failure of the sheet metal mouse barrier around the stack to serve its intended purpose. Either because of small cracks where the sheets overlapped, or the ability of mice to dig burrows underneath, the haystack was seemingly always well stocked with mice, and as a consequence of that, equally well endowed with cats, (the 'shed' cats) who sized up the presence of the truck as an indication that lots of mice were about to be disturbed and revealed. It was quite a scene, with cats, and kids, chasing mice all over the hay while the loading took place!

Then the chaff-cutting got underway, with Dad carrying the sheaves from the truck over to the chaff cutter, dumping them in the feed-in trough on the machine, and cutting the band of twine with his pocketknife, so that Fafa could feed the hay into the jaws of those evil looking cogs!

That's another memory of the farm – there was always plenty of binder twine tied up into bunches, hanging on nails in all the sheds, having been cut off sheaves and bales of hay. If you wanted to tie something up, you could always find a piece of string.

The final fun part of the chaff-cutting was inside the Chaff House, which was windowless, and only had one small entrance door. The good part of how this was used, was that against the far wall, Dad and Fafa used to build a small but quite high stack of 'square' baled hay. The square bales were made by a different machine to the binder, owned by a neighbour, Sandy Morrison. They were much denser and stacked solidly together. A perfect launching platform from about four metres up, from which to jump gleefully into the incoming heap of light, fluffy, and very dusty, chaff, as the elevator tossed it in. A bit like jumping into a giant pile of feathers!

From deep within the heap, you could wriggle back to the surface, lie there and watch the beams of sunlight coming in through the many nail holes in the walls and roof, like search-light beams in all the dust. The only downside to this fun was how itchy you became, and how much dust got up your nose!

Sunday Drives

A linkage between haystacks and going to church on Sunday might seem a bit unlikely, but stick with me, it is a strong childhood memory.

A further bit of scene setting: Firstly, the fifties were a series of excellent years for rainfall, so the crops grew well. And also, the crop varieties available to grow at the time grew much taller than modern varieties of wheat, oats and barley. Thirdly, because the harvesting machines of the day were not 'Auto Headers' but were Power Take Off (PTO) machines towed along by a tractor, a space between the fence and the crop was required to enable the tractor to be driven around the paddock on 'the first round' of harvesting without driving over and flattening the crop.

The solution to this issue was simple and a very practical one, given the need to ensure there was always a supply of fodder in the form of hay on the farm. This involved 'Cutting the Tracks' around every crop with the Reaper Binder, or a mower followed by a baler, and then carting the hay away on the Bedford truck to be stacked. This gave a clear run for the tractor right around the paddock at harvest time. Often, in fact mostly, the hay 'tracks' weren't just cut around the paddock edge beside the fence line, but also were cut through the middle of the crop, dividing the paddock into 'Lands'. This generally was a divide between different wheat varieties planted in the same paddock. And you know, the 'Hay Tracks' just made a paddock of wheat look good – with nice, neat edges to it. Hold that vision.

Going to church on Sunday morning is just what you did in the fifties in a rural district. It wasn't particularly open for discussion or choice, you just did. So, every Sunday morning Mum would dress us kids in our Sunday best, put on her best outfit, while Dad put on his suit, and ensured that us boys had our ties tied, and hair combed, just as neatly and correctly as his, while Shirley had her best dress properly fitted.

Then we all squashed into the Holden Car, and off we went. Dad always was a Holden Man. He was proud to claim he bought the first Holden (A cream 'FX") to come to Nhill, and always traded in on the next model every two years or so. We rode around in many Holden's through the years, never anything else. That you did so was almost as strongly held a belief as it was that Collingwood was the only possible football team one could support. Both of these were bedrock articles of faith, I am afraid to relate, held with somewhat more conviction than that which saw us driving off to church.

In the early fifties, the trip was just a mile up the Perenna Hill and down again to the mudbrick Methodist Church that was beside the school building. But as the faithful dwindled, or maybe it was that the old structure was no longer suitable, and anyway, the road to Netherby was now mostly metal with only a few dirt sections, our place of worship became the Methodist Church at Netherby.

Sunday mornings in Netherby were really quite an occasion. There wasn't just a Methodist Church in Netherby; there was a Lutheran Church, a Church of England, and a Catholic Church as well. So, all the roads to Netherby were busy, before and after church.

It was only later on in my time that I began to realise what an interesting part of the district's development and heritage this really represented. The district was settled by mainly second-generation migrant families of European descent, and they brought their ethnic and religious traditions with them. And there was diversity. There were the Irish families (Catholic), Brits and Scots (Methodist and Church of England), and Germans (Lutheran) This ethnic diversity was most expressed by Sunday morning church goers; for the rest of the week, the district (at least it seemed to me) fell into a comfortable common set of shared beliefs and interests – When was it going to rain again? It was a pity the footy team lost to Yanac yesterday! When are they going to fix the road to Nhill?

Thus, we drove up the hill into Netherby in our Holden, past the Pub, Deckerts General Store, and the Hall (or Mechanics Institute) on the left, and the Fire Shed and the Motor Garage with big wooden doors on the right and turned the corner into Church Street (that also lead to the footy ground). There was something of note at the Motor Garage that I might come back to later. The first building on the left was the large and imposing brick structure of the Lutheran Church. This was the largest congregation, so the cars parked here spread around the corner up the main street. And they were good cars! You see, the German origin families who had mostly come across the ninety-mile desert from South Australia, as the good land there became closely occupied. They were good farmers, and being mostly first on the scene, occupied most of the good, heavy 'Buloke' country in the district, and grew the best crops. Hence, there were innumerable Ford Customlines, Chevrolets and Chryslers parked in front of the Lutheran Church! (no Benz's that I can recall)

Further up the street was the much smaller wooden Church of England. Parked in front of this place of worship was a more modest fleet of vehicles – Morris Majors, Hillmans, the odd Ford Zephyr and a few Holdens. Right in front was the Vicar's Humber.

Next door was our Methodist Church. This car park was Holdens and Zephyrs, and a few Morris, but all quite new, to which our latest model Holden was added.

And away around the corner was the Catholic Church, with a more eclectic collection of vehicles, from Chrysler Royals to pre-war 'A Model' Fords.

As I look back, I realise my memories of Sunday Mornings in Netherby were a really valuable window into a better understanding of how that district was settled and the values and beliefs brought to the area by the families who settled there.

It was an early example of a multi-cultural community, or at least, multi-European culture. It was a successful community of what I would describe now as Yeoman Farming families, successful in adapting European practices to this corner of the Wimmera.

What about my memories of being raised in the fear of God? I can still remember some time in my early teens fainting and collapsing in the Pew as I pondered the sermon of a particularly zealous and strict minister. He had me believing I really was off to Hell, and without passing Go! He scared the wits out of me! I really hope no kids are subjected to this kind of religious indoctrination nowadays. I look back on it now as a form child abuse.

But it wasn't all bad. As I grew, I used to really enjoy the times I was asked to deliver a reading. Even though I often didn't understand, or didn't believe what the text was saying, I did find the experience of reading to an adult audience a valuable exercise – an early introduction to the art of public speaking which I am sure benefited me in later life.

It is probably important to note here, in case it doesn't become apparent in the rest of this narrative, that social norms back then were very different to today with respect to religion, race, and equal opportunity.

For starters, a centrepiece of Government Policy was the 'White Australia Policy', which was finally repealed in the early 1970s. This policy strongly regulated the nation's immigration policy, and therefore the make-up of society, which when I was young, was almost entirely white European in origin. Religion was also a central aspect of how society was organised, and because of our racial background, was dominated by people of Judeo-Christian belief, broken into different groupings as the several churches in Netherby attested.

And I was also brought up at a time when men dominated every aspect of life. Women had the right to vote, but in just about every other aspect were treated as less equal than men. Much has changed since about core beliefs and attitudes which shaped society since – I hope I too have changed and adapted accordingly!

Probably the best aspect of my religious upbringing was the Methodist Youth Fellowship. (MYF) Now there is a whole new part of my story. The MYF is how I came to meet Mary. The MYF was a rather informal group of young people of the faith (or ethnicity), mirrored by the larger local 'Luther League' for young people of the Lutheran faith (German ethnicity).

With the Lorquon and or Netherby Sunday School Hall as a venue, it was an early introduction to boys and girls, or rather young men and women, getting together socially. In our case with the aid of the modern wonder of a portable record player. I can still picture the player we mostly used, supplied by Kelvin Deckert. It was all packed up in a special grey case, like a suitcase, and we keenly assisted Kelvin to get it out, lid unfolded, arm and needle removed from its protective holder and hoping that the batteries had enough charge to see the evening out, as we sorted through which EP's and LP's we wanted to play.

With the aid of the music, we enjoyed a range of square dancing, then venturing into ballroom dancing, interspersed with games such as 'spin the bottle' and 'pass the parcel'. They were good times – until the arrival of a new parish minister, whose view was that any form of waltzing or modern dancing was way too high a moral risk to be countenanced as an activity suitable for MYF! He must have seen himself as the last bulwark against the tide of the new world of Rock n Roll, of Elvis and the Beatles that was sweeping the globe at the time.

Travelling quite some distance for social activities was never really an issue for us back in the 1960s – we would think nothing of driving the one hundred kms to Horsham for a night out. That was only an hour away – or less! There were no speed limits, other than being potentially booked for dangerous driving, so it was quite normal to wind the Holden up to 75-80 MPH (130-140Kmh) down the highway and be there in no time! Not only that, but they had dances there where you could dance the night away without the disapproving presence of a straight-laced man of the cloth – and call into a pub to have a drink or two as well. And it was at a Horsham MYF dance where I met Mary, in 1967.

Footnote to speeding when driving: It is also a fact that more than one close acquaintance never made it to an age like me with an ability to write a memoir about the last seventy years. They died in crashes up against trees or trucks or something else, with speed and unsafe cars the cause.

But back to Church on Sundays and haystacks.

So now I'm back to one of the good parts of our Sunday excursion to church and back, when harvest time was approaching. All those 'hay tracks' cut in the crops were too alluring for Dad to ignore. "What about having a drive around the crops?", he would say, as he turned into an open gate on the road home. He never had any opposition to that idea, particularly if you were one of the kids with a back window seat in the Holden Car.

Off we would go around one wheat crop, up through the middle of another, with Dad happily saying, "Look at that Pinnacle Wheat!" " Geez, it's a good variety!" "I reckon that will go twelve bags to the acre!" The heads of wheat would be flying past at window level, and you could stick your hand out and feel the heads going 'thwack thwack thwack' as the car sped along.

"Let's go and have a look at the Olympic in the Front Paddock". 'Olympic' was the new breed of wheat released in 1955, just before the Olympic Games in Melbourne. This was a quite tall growing variety, so it was like driving in a tunnel – we could barely see over the top of the crop.

Those drives around the crops just before harvest in a good year were a highlight of life on the farm and having those 'Hay Tracks' cut around and through them made it special.

And as my farming years went by, I have to say, some of the best times were out walking into a maturing crop to check its progress. There is something deeply satisfying in standing out in a waving field of wheat or barley that you have grown, gazing across it to a distant line of Mallee trees, and the horizon beyond.

We enjoyed our Sunday Drives!

And perhaps as a footnote, I have one other recollection of our Sunday drive to Church. Most, but not all locals were devout attendees as we were. There were a few local characters (names withheld!) whose Sunday morning was not at church but sitting on empty kerosine drums out in front of the motor garage, enjoying a beer or two together as they watched the district pass by. I recall the disdain with which this was held, and perhaps pity and concern for their souls:" Look at them – at their 'Fox Hole' like they always are".

Were they lost souls? Or just ahead of their time? Maybe they were Netherby's precursor to today's Sunday morning ritual at Cafes in every town and city across the land.

Rainy Nights

We grew good crops we could drive around on Sunday afternoons because the 1950s and 60s were good wet years in the northwest Wimmera. One of my earliest memories, probably around the year 1950, was of long, rainy nights through the wintertime, the sound of rain pounding down on the tin roof of the farmhouse, which had no such thing as insulation back then. You could actually see outside through cracks in the wall! So, you could hear the noise of the rain, hear the wind and feel the cold, wrapped up in bed covered in heavy woollen blankets and eiderdowns. The sound I remember most was the banging against the wall of the old canvas blinds that were the only barrier to the elements on the veranda and 'sleep-outs' on the south and west side of the house. I can still hear those blinds thrashing about as the bursts of wind and rain and showers kept up through the night! And the sound of the spouts overflowing, with the water pouring down onto the outside wooden veranda floor and footpath.

Don't get me wrong – this wasn't bad or sad. It was great! You would wake up in the morning, and if the rain had cleared and the wind settled, the next really memorable sound was that of all the crickets and frogs, rejoicing in their newly rain-soaked environment. There is something really special about waking up to a still morning after rain in the Mallee – those particular sounds are unique to that landscape. If you can manage it, find an opportunity to be there, and wake up after a rainy night in the Mallee and witness what I am talking about – it is a special experience, a lasting memory of mine. An experience that is now all too uncommon, particularly as our winter rains are diminishing due to climate change. How lucky were we that we lived as kids through a particularly wet period of years that were somewhat abnormal then, and not really repeated since, regardless of the subsequent reality of our changing climate.

On waking up, the next sound was that of Dad (Cliff), tramping along the old wooden front veranda with its loosened, creaking boards, on his way to his early morning ablutions. Then back to roll up the blinds in the bedroom to let the light in and saying "Look out at the front paddock! The crabholes are all full again! What a beautiful rain! I'll go and measure it! Must have been nearly an inch!"

Off he would go, and we would peer out to see that he was right – there was water everywhere across the bare fallowed ground. That wasn't just because we had had lots of rain, but also because of the farming methods of the time, which involved cultivating and keeping the land bare for the previous twelve months prior to sowing the new wheat crop. The finely worked shallow topsoil would shed water very quickly, but that's an issue we might come back to later.

Dad would head off towards that all important fixture on a fence post out the back of the house – the rain gauge, but not before diverting to the little tin shed that housed the thirty-two-volt power plant. Soon we would have another sound to go with the crickets and frogs – the 'Pop-Pop-Popping' of the slow revving Cooper engine as it burst into action, enabling us to pull on the string hanging from the ceiling and turn on the electric light, as Mum (Margaret) was doing the same in the kitchen, before filling up the firebox of the old black stove, around which us kids could soon huddle for a little bit of warmth as the kettle was set to boil, and the pot of porridge stirred.

Back inside would come Dad, with the glass jam-jar out of the rain gauge, getting down from the mantle-piece the actual glass measure which for some reason didn't reside in the rain gauge, and then slowly pouring the water from one to the other before holding the gauge up and saying – "See – seventy-two points! I told you it was nearly an inch!" (An inch of rain equalled one hundred points) He would then proceed through the door into the passageway that ran down the centre of the house, where he had the rainfall chart pinned up beside the wall mounted telephone. The seventy-two points would be pencilled into the correct date on the chart, and the next thing we would hear would be the winding of the handle on the old party line phone. Dad loved talking on the phone! He would be ringing all the neighbours, to see how much rain they had measured. "Is that you Berly? (Eric Berlund) How much rain did you have? We had seventy-two points!"

We would be all finished with our breakfast of porridge and toast, which was produced by holding it on a big wire toasting fork in front of the open door on the wood stove's firebox, because I don't think a thirty-two-volt electric toaster was ever invented, by the time that Dad came back into the kitchen. He was able to inform us that Berlund's had only had sixty-five points "out there" (They lived about five miles further north, near the edge of uncleared scrub of the Big Desert, separated from the 'good' country by the Vermin Proof Fence), but that Sandy Morrison had seventy-five Points, and Evan Evans had a hundred and two!

There was nothing really unusual about this focus on rain. In a district where the long-term average rainfall is less than four hundred millimetres (about fifteen inches on the imperial scale), the amount and timing of rainfall was, and is increasingly today, the most critical influence on everyone's livelihood.

I think the importance of this recollection is to underline our good fortune in time. The 1950s, and the early 1960s were, as the rainfall records show, one of the most consistently wet periods of time during the past one hundred plus years of farming in the region.

They were good years to be farming – always enough rain for the crops and pasture – sometimes too much, allied with good prices during the post war period where food and fibre was in shortage, and in high demand. Plus, the government guaranteed a minimum price for wheat.

As a result, my reflection is that we had a fortunately comfortable existence as measured against the overall populace at the time.

And we lived, as mentioned before, in a quite vibrant and active little district. It was, as I recall it, a generally cohesive, inclusive society, free of the class distinction that typified the story of Western District settlement. It wasn't my recollection that there were any families in the area who flaunted inherited wealth, or in the thriving business community in Nhill. The generally good standard of living was as a result of their own endeavour.

Down the Timber

Figure 8 The Timber Paddock (Photo by Ian Oswald Jacobs)

Those rainy nights transformed our larger playground - the 'timber paddock'.

The twenty-acre (eight hectare) timber paddock was pretty much the only part of our then twelve hundred- and eighty-acre (two square miles) farm that was still in its pre-cleared original state. Our house, plus the farm buildings, were located on the northern end of this patch of remnant vegetation, the centre of which was somewhat lower than the surrounding land, and as such an area to which water would naturally drain, if sufficient rain fell. In the parlance of the times, it could be described as a 'swamp'. The evidence is that, prior to European settlement, it was probably a fairly unusual event for this whole area to flood to any depth – but most winters would have seen the filling with water of several low lying depressions in the centre of the timber – which us kids referred to as 'the ponds' – no doubt influenced by Dad (Cliff) who in fact had lived here since the house was built in 1929.

In our families experience here– the 'Timber Paddock' has in fact flooded to full depth of about four metres before it overflowed further to the northwest twice –

in 1930, and again in 1974. On both occasions, the water at its peak was lapping under the floorboards of the southern side of the house. It seems a reasonable assumption that these were rare, perhaps in recent history unprecedented floods – given that the area was referred to as 'Pine Flat' at the turn of the twentieth century when Jack Dodds arrived there. This was due to the fact that the apparently dominant tree was in fact the native pine (Callitris species)– which do not grow in areas subject to inundation. I can clearly recall as a child there being lots of stumps of these trees throughout even the lower areas of the 'timber', which had all been cut down and used for building materials for the various farm buildings. The straw roofed 'Cow Shed' which was what remained of the earlier more extensive Horse Stables, the 'Chaff House', the haystack stands, the 'Pig Sty', the stumps for the Barn, and more.

The flooding of 'Pine Flat' on the two occasions recorded was most likely due to the change of land use when the land was developed for farming – wholesale clearing, and the use of bare cultivation or fallowing, which rendered the land much more likely to the 'run off' of water rather than it being able to more slowly infiltrate downwards.

Back to our 'crabholes' in the paddocks, and 'ponds' in the timber, and the fun and learning we had there. This spanned generations, as our own kids played and learnt in that same area in the same way that I and my siblings had, and I presume my father and his sister beforehand. Back to those wet winter mornings...

It's mainly a recollection of gumboots full of water, and very wet socks! But what fun it was to head off down from the house to see how many of the ponds had filled overnight. And to walk into them to check the depth!

They were never particularly deep, in fact generally about one inch shallower than the height of a kid's gumboot, unless you were unlucky enough (or lucky, depending on your perspective) to step in a 'runaway hole'. A runaway hole was a small area that would have cracked open during the long dry summer, and then subsided into a deeper hole when inundated again.

But apart from sloshing around and getting full gumboots, I am sure that this was where I first started to understand the amazing natural world, of which the 'Timber Paddock' was, and hopefully still is, an example of what was once so much more extensive. Because these little ponds soon burst forth with life – tadpoles galore, but also other rather amazing little water creatures such as 'fairy shrimps', and much more. It really was a natural wetland 'museum' at our back door.

It also provided an introduction to engineering. As the rainy nights (and days) continued, the several ponds would fill and overflow to the next one, so we began, with a shovel, to intervene and speed this process – digging little drains from one to another, and then watch the results. Like playing at the beach. The Timber Paddock was a kid's paradise!

On the subject of digging, there was another fun activity down the timber, which involved some other willing participants – our dogs. Given ours was a mixed farm with livestock to manage, sheep dogs had a very important and necessary role, and so over the years several good and faithful dogs came and went, occupying a core role – involving both work and as fun companions.

Therefore, I have good memories of some of these dogs, including the first one I really recall – Ginger! He was a classic big hairy Border Collie. I'm not sure about Ginger's competence as a sheep dog, but boy did he like digging out rabbits! Now it happened that around the western side of the timber paddock there was a bank of sandy loamy soil, which provided a perfect location for rabbits, which were rather abundant at the time, to dig their warrens. This was prior to the advent of myxomatosis, which greatly reduced the wild rabbit population.

We spent many hours as kids down there with a shovel, helping Ginger shift tonnes of soil, as he dug along the various branches of rabbit holes, which were often a half a metre or more below the surface. He would dig till he puffed like crazy, took a break to sniff for scent to see if he was pursuing the right branch, trot over to dunk himself in any water he could find, and then resume digging. I think us kids and Ginger made a bigger contribution to soil erosion than the rabbits did.

All in all, the timber paddock, and the immediate environs of the farmyard itself were a wonderful place for kids to grow up in. There was so much to do and explore. Like lifting up old sheets of iron that seemed to be scattered around the place and seeing what was there. Almost certainly some kind of lizard or skink, heaps of spiders, lots of mice, and very occasionally a snake. A real-life nature lesson.

For me, there were so many options to improve my football kicking skill! Like over the machinery shed, through the half open barn door for accuracy, and after the little window at the top of the barn's end wall fell out, through the little square hole it left behind. It took a good torpedo punt to go through there from twenty metres back! As to whether years of this was why the barn door hinges were rather bent and loose, and the corrugated wall at the end looked a bit dented and loose is another matter.

Another aspect of this area which didn't gel with me until much later in life was the evidence of its use by Aboriginal people long before our arrival. The evidence was the collection of stone tools that from time to time we unearthed in the general vicinity – stone axes and grinding stones. These were apparently found quite accidentally in the decades before, and during my time, mostly due to some soil disturbance such as cultivation in the paddocks immediately east and south of the timber. I guess they were kept by the family as curios, but there was never really much discussion about how they got there, or what they represented. I have some recall of Dad talking about at least one aboriginal man whom he knew, and who must have had some connection with the area, but no more than that.

Much more recently I have read of the fairly sketchy knowledge of pre-European occupation of the area, mostly pointing to the established fact that there was quite a community based along the Wimmera River, and Lake Hindmarsh to the east. This suggests that these people would often venture in the winter and spring months away from the environs of the river and lake, taking advantage of the seasonal wetlands to the west. Thus, we can only wonder about how frequently people came to camp by and enjoy the Timber Paddock and its ponds, as we had enjoyed it. The tools left behind attest to the fact that they had.

When we left Perenna in 2000, Mary packed up all of these relics in our possession and gave them over to the Barenji Gadgin people in Horsham for their safekeeping. I am glad she did so.

I never lost my appreciation of this natural area over the fifty plus years I lived there, nor did Mary, nor our kids as well. We have often reflected that as time and events unfolded, the broad expanse of the open farmland came to hold less emotional importance to us, but the timber paddock was always what we particularly valued.

Beyond the timber paddock, the rest of the farm was cleared land – in fact very well cleared land! The Dufty's, followed by Jack and Cliff did a rather complete job of removing the Mallee vegetation that originally covered the 1280-acre farm at Perenna, surrounding 'Pine Flat'. Apart from a thin line of Mallee left along the mile long northern boundary, very few trees remained – in fact, in later years – post the 1982 drought, which we will get to later, this became somewhat of a focus of mine, at which time I counted the number of original trees to be less than a hundred on the whole two square miles. But what it meant, was that the farm was very well set up for agricultural use, with nothing to impede cropping operations.

Living off the Land

The farm provided the basis for our diet. We ate our way through heaps of eggs thanks to the productive little 'chook house' over behind the barn, and the occasional roast chook, (never referred to as roast chicken!) We ate heaps and heaps of mutton, and this was supplemented by lots of milk, cream and butter all thanks to the two cows that faithfully came into the "cowshed' every morning and night to be fed and milked.

The straw roofed cow shed was 'down the timber', about 100 metres from the house, with two or three haystack sites adjacent, plus the corrugated iron 'chaff house'. That old cow shed was a typical example of the early farm buildings, which have now mostly disappeared from the region's landscape. They were constructed almost entirely of straight long logs cut from the local vegetation as the land was cleared. The native pine (Callitris species) was ideal, being relatively light and straight, and were used in the walls, stood up vertically, between larger posts used for the building frame. The posts which formed the edge and gable line were the tallest gum (Eucalypt) trees that could be found, ideally with a good fork at the right height to hold the horizontal beams, on which more poles were then laid, all tied together by wire, to form the roof. The roof poles were then covered with a thick layer or 'thatch' of straw, another of Fafa's skills being to periodically add more straw to keep it entirely waterproof.

The cow shed was apparently the remaining remnant of a much larger couple of straw sheds which were used as stables for the farm's horses. I have no recollection of these – they were dismantled after the horses were replaced by tractors; the timber being used as a source of firewood.

Milking the cows was a seven days per week, three hundred and sixty-four days of the year job – you might say central to our existence, so I do retain strong memories. There were two 'bails' with feeder troughs for the two cows. The first task was to take a couple of big old square kerosene tin buckets to the chaff house, where you scooped them full of chaff, plus a smaller bucket into which you scooped crushed oats which were in open jute bags just inside the chaff house door.

Back to the cow shed to top-up the cow's feeders, which they were always happy to see, diverting their attention just enough to allow you to place a leg rope around one hind leg, and tying this to one of the wall posts. This was to stop the cow from kicking the bucket over, and also to stop the cow from kicking the milker over! Dad, and later us kids, would then sit down on an upturned kerosene tin with a bag draped over it for cushioning, and set about milking each cow in turn.

It wasn't just us kids that found this as a fascinating activity – it soon also drew the attention of our pet cat of the time, who had followed us down from the house, but also two or three semi wild – or fully wild – 'shed cats'. I didn't think much about these wild or feral cats at the time – the justification being that they were there to 'control the mice'. I am sure they did control some mice, although there always seemed to be plenty left over, but also as we have subsequently come to understand, the spread of domestic cats into the wild has had a massive impact on our native animals and birds, and our farm helped to speed that unfortunate process.

The loss of native animals and birds wasn't seen as an issue – they just all came under the general heading of 'Vermin' – anything that moved or flew and wasn't a farm animal was in this category. If any unfortunate bird or animal was in range, you just shot it if you had a gun handy without thinking much about it. You got rid of it. That was how Jack Dodds understood his role in the landscape, I think Dad (and Mum) saw it a bit less that way. This was the time in which I was brought up. Nature was seen as a huge and inexhaustible thing, largely a challenge to be subdued in order for us to prosper. It was only the passage of time that revealed how wrong this was, and that in fact what was happening across the whole Australian landscape was the slow decline, and often extinction of plants and animals.

And a good example of this slow realisation that you never valued something until it was gone was played out at Perenna – the disappearance of Bush Stone Curlews. A very early recollection, particularly remembering those rainy nights, was how Dad would often come in from the back door and say, "Listen to the Curlews calling down the back – that's a good sign of rain" And we would go out and listen, and sure enough, away in the distance, you could hear the haunting call of these birds. It was a slow realisation of mine, that as time went on, we never heard this particular bird call anymore. I'm positive, that a generation later, our kids never ever heard a Bush Stone Curlew at Perenna. They were gone. These birds are still seen in good numbers in northern Australia, but are now rare in Victoria, and have mostly disappeared from the Wimmera. Their nests being on the ground had been fine over the millennia, until us Europeans came along with our Foxes and Cats, and the ability of the Curlews' offspring to survive until maturity was greatly impacted.

Back to the Cow Shed. The cats did enjoy milking time, started off by our pet cat being game enough to get just in range to be squirted in the face with milk, straight from the cow's teat! This of course was a discomfort worth putting up with for the likely result of several squirts straight in the mouth and down the throat. Seeing this highly beneficial outcome, the wild cats would creep slightly closer from underneath the cow's feeder trough, where they would cop a face full, jumping back, before again venturing forward for another go. It seemed there was plenty to go around. We would end up with a couple of buckets-full of milk, while the cats spent the rest of the morning sitting in the haystack licking themselves clean.

Meanwhile, we would head back up to the house to separate the milk into cream and skim milk.

This took place in the 'separator room', which was a closed off part of the veranda at the back of the house. The separator was actually quite a clever machine, which used centrifugal force to spin off the cream from the milk.

It had a big bowl on top, into which you poured a bucket full of milk fresh from the cowshed, below which the circular separator mechanism sat, with two spouts placed so that the cream and separated or skim milk was directed into a couple cans placed on the bench beside the machine.

On the side of the machine was a big handle that we had to turn to operate the machine. I never did get around to pulling the old separator apart even after we ceased using it to see how it worked, but what I can remember is how hard it was to get it to begin to spin in the first place. It must have had a really heavy fly-wheel inside it. Anyhow, we would get hold of it by both hands, and with a big effort get it to slowly start to spin. As it slowly built-up speed you could hear the works starting to 'whirr' as they spun round and round, and in particular you could hear the little bell on the back of the machine begin to make a slow, dull 'click-click-click' sound. As you turned the handle faster and faster, the bell's sound began to change to a higher pitch and faster frequency, the machine's whirring sound built up, until finally the 'click-click-click' became a 'ting-ting-ting'. This was your indication that the machine was now up to the correct speed, at which time you would turn on the tap beneath the bowl on top and allow the milk to slowly pour into the centre of the centrifuge, and the separation process would begin. Then your task was to keep turning that handle fast enough to make the bell keep on 'ting-ting-tinging' and not 'click-click-clicking'. Otherwise, all of the milk would go straight through to the skim milk container with no cream produced. If you wanted cream, you kept up the pace!

Therefore, we always had plenty of cream, which we poured over lots of bread and jam, and heaps of puddings – chocolate pudding, 'roly poly' pudding, golden syrup pudding, and much more. The balance of it was made into butter for our own use, using another hand-operated device, the butter churn. That was probably a harder physical test than the Separator. This was a bowl with fixed paddles inside, that rotated as you wound the handle, having placed the cream and a heap of salt in it. You turned and turned, changed from right to left hand to give your arm a rest, kept turning till that arm was aching, then back to right hand, and turned and turned and turned! It seemed to take forever, but finally you ended up with a mass of butter, separated out to leave what we called 'whey' or butter milk.

But we didn't consume all of our cream ourselves. Most of the cream in fact used to be poured into a big metal 'Cream Can' kept under the bench in the separator room. It would take several days to fill this, which when full, would be put in the back of the Ute, and off up to Netherby, to be put on the weekly goods train and sent off to the Rainbow Butter Factory, exchanging it for the empty can sent back out on the train from last week's delivery.

What we did consume a lot of was meat. In particular, sheep. Nowadays it's mainly referred to as lamb, but that was seldom a choice for us. We ate mutton; lots of mutton. Mutton is actually the meat from a sheep somewhat older than a lamb, often several years older, and therefore bigger, stronger flavoured, and tougher.

I'm not really sure why we ate mutton instead of lamb, because we did breed the animals, and they did start off as lambs! I suspect it may have been because slaughtering and butchering an animal was not the most pleasant task, so if you were going to do it, you might as well make it a big old one and get plenty of meat as a reward for your effort. Killing a sheep, or a chook or the odd turkey was of course just a part of life on the farm. There was nothing unusual in coming home from school and hanging around and watching Dad or Fafa slaughter a sheep, and see it placed in a white calico 'meat bag' and hauled up beneath a black box tree using a pulley and rope, to be allowed to 'set' overnight before the task of cutting it up into the various joints of meat the next morning. I reckon we ate close to a sheep a week! I am sure this explains why I am not a particular fan of roast legs of lamb or big forequarter chops nowadays, as we pretty much ate a lifetime's worth of lamb (or mutton) back then.

We were not so independent when it came to fruit and vegetables. There were some fruit trees around the house at Perenna, and a vegetable patch, but it was never particularly productive. Maybe this was due to a lack of commitment, but it was largely due to the use of bore water with its high mineral (or salt) content which meant that no matter how much water was applied through the hot summer months, the plants did not thrive as they may have if rainwater was available.

This didn't mean we went without this important part of our diet. There was a regular trip to Nhill for supplies, generally of a Friday, where an ample stock of fruit and vegetables in season were available at the Greengrocers, the main one being that attached to the 'Central' café. I believe, unlike today, that at least a portion of this supply of fruit and vegetables was produced within the Wimmera, such as at Dimboola where a Chinese family had a long-established market garden.

I do have to relate a memory here that might be better left unwritten, but it does indicate old attitudes and beliefs that were part of our experience and influenced us at the time - which now look so wrong.

Dad turned up home from Nhill one night and produced a brown paper bag full of some unusual looking things. He proudly said, "Look what I bought today! 'Chinese Gooseberries'" These were actually what we today call Kiwi Fruit. Now, we had never set eyes on these before, and I can remember being a bit doubtful as he invited us to "Have a taste of these" "They're delicious!" as he bit into one, showing us the half-eaten inside of it which looked unlike any other fruit I'd ever tried. Just then, Nanna turned up in the kitchen to see what was on offer, and in a disapproving way said (Racist Remark Alert)

"Cliff, why ever did you bring those home for these children?" "You know they've been grown by those dirty Chinamen!" I think that under the cover of the apparent moral authority so provided, I chose a banana instead and lost the opportunity to taste my first ever Kiwi Fruit. Multi-culturism was not part of Nanna's world! Dad continued to enjoy his bag of Chinese Gooseberries by himself.

There was a lot of good in Nanna: she was well meaning, and very kind and generous to us. But she was definitely a product of an earlier less enlightened time. A time when the White Australia Policy was the law of the land, and Aboriginal people were not even counted in the census.

It was also on the calendar that Dad would head off down to either Riverside or Quantong near Horsham when it was time to harvest peaches, plumbs and apricots, and come home with several crates of freshly picked fruit. It would be Mum and Nanna's job for the next week to peel and slice all of these and place them in big glass jars that were all in turn processed in the 'Fowlers Vacola' preserving unit, placed on top of the wood stove. Once cooked for the prescribed time and temperature, steel lids with rubber seals and metal clips were placed on these, which as they cooled created a vacuum seal to ensure a good supply of preserved fruit.

The other food stuff that quite often turned up in bulk was potatoes. It seemed that almost every time Dad returned home from a trip to Melbourne in the car, he would have a big bag of unwashed potatoes in the boot. He made it a habit to pull into a grower's shed beside the Western Highway in either Bacchus Marsh or Gordon and purchase a bag of potatoes. This was a big three-bushel jute bag, the same as we used to fill with wheat at harvest time. That always kept us in spuds for a while!

Home cooking was also a key part of life on the farm – lots of cakes, pastries, biscuits and puddings. This all required lots of flour and sugar, so that too was stored in volume. There were two big metal bins on the enclosed back veranda, with hinged lids. These were regularly filled with bags of flour and sugar – and not just kilogram bags. More like Twenty-, or Sixty-Pound bags as per the imperial measurement of the time.

But we didn't have home cooked bread – because we had fresh baked bread, straight from the bakery in Nhill, delivered to our door by 10.30 every morning. This happened by courtesy of our 'Mailman'.

The Mailman

The legendary reliability of mail services was indeed true of our daily lives out thirty kilometres north of Nhill. The Mailman's arrival every day of the week could be relied upon, so much so that that was how we had access to fresh baked bread every morning, picked up by the mailman from the Bakery in Nhill, along with all of the mail. This was pretty impressive when you consider that much of the mail route from Nhill to Netherby and Lorquon and back was not even a gravel road let alone bitumen. Much of it was dirt, which was trafficable most of the time, but a challenge after rain, particularly back in those wet years of the fifties. But everyone watched out for the mail man, and if there had been lots of rain along his route, there would always be a tractor and chain ready for the rescue if he had slipped off onto the side of the road and became bogged.

It was part of the rhythm of life that the Mail Cart drove in our front gate at around 10 o'clock every morning, (unless he had been bogged) and personally dropped off our Private Mail Bag, our Bakery delivery, any parcels we might have been expecting, and a whole stack of newspapers.

This personal delivery was one of the benefits we had due to our house being right on the road, making his drive into our back door easy and quick. And a chance for a quick yarn about things before he hurried off on his run!

Most farmhouses were set back from the road, so they had a mailbox installed at their farm gateway, of sufficient size to accommodate everything that might arrive. These weren't tiny little letter boxes! The most common type was a forty-four-gallon oil drum with its end cut out and reattached with a hinge to make a door, but there were also lots of ex war metal Ammo Boxes, and then some more creative wooden structures. Most had the family or property name painted on, plus the Private Mail Bag number.

Pretty much everyone had a private mail bag – which became your postal address. Ours was:

Private Bag 34
Nhill
Victoria

There was no such thing as a postcode back then.

The Private mail bag was made of canvas and had a leather strap and buckle attached so that the top of the bag could be closed and secured. Your private bag number was stencilled on in black letters and numbers. We actually had two identical bags, which were supplied to you by the PMG (Postmaster's General Office) for an annual fee. They were swapped each day when the mailman arrived.

The daily arrival of the mailman is a strong memory of my childhood. It was a quick 'good day' to the mail man, as he passed out of the window of his vehicle the mail bag, the loaves of bread, and the heap of newspapers. We would in turn hand him the second mail bag – into which we would have placed any letters we might have had to post, and he would be off on his way.

Then it would be into the kitchen, where the mail bag would be opened, and its contents tipped on the table. The next hour would be easily spoken for, while we had morning tea, often including fresh bread and butter, while the mail was opened, and the newspapers read.

And we did get a heap of newspapers. There would be three Melbourne dailies – 'The Sun, 'The Age' and 'The Argus', which editorially covered off most political viewpoints of the day. And all of which had been transported up from Melbourne to Nhill overnight. Then there were the weekly papers which would come on various days – The Nhill Free Press, the Weekly Times, and the 'Countryman'. I do reflect that one of Dad's good influences was ensuring there was a plethora of material for us to read. I'm sure this was one of Mum's influences too – as every trip to Nhill usually meant that there would be more books, magazines or comic books to be read.

One of the paradoxes that puzzle me is that despite all of the change and advancement we have seen, there are some services that were more available, faster, and or better in the 1950s and 1960s than they are today. The daily mail was a good example.

Other examples were the regular passenger air service from Nhill to Melbourne, flying in a 'DC 3', and the 'Day Train', which was a train from Serviceton on the South Australian Border to Melbourne, and one the other way back up from Melbourne. Add to that was the Overland Express – one train each way overnight between Adelaide and Melbourne.

And if you decided to drive, the Western Highway was sealed all the way to both cities. (I think the last section to be sealed was completed in 1955 - between Bordertown and Tailem Bend). With much less traffic, and no speed limits beyond town boundaries, the drive from Nhill to Melbourne was quite possible in four hours. Hardly possible today.

A little bit more about the farmhouse. The main living area was referred to as either the "Dining Room or the "Lounge Room." A better description would have been the Living Room, because its where we spent most of our time indoors. It had a corner fireplace, but the highlights were two key items. These were the Large Dining Table in the middle of the room, draped by a big maroon or red cover, edged with golden tassels when not in use, and an upright Piano set against the wall. There was more to both of these items of furniture which made them of great value and importance to the way we entertained ourselves.

The dining table was a very solidly made, quite heavy piece of furniture - it actually was a small 'quarter size' billiard table. Its polished timber top was made in three movable pieces, which, when removed, revealed a slate based billiard surface. A billiard cue rack was another item in the room, mounted on the same wall as the Piano.

We played lots of billiards, both amongst ourselves, and as an added entertainment when we had visitors. Billiards was a very popular entertainment back then, all the more so because of the famous Sir Walter Lindrum – Australia's own world champion Billiard Player. He was almost as highly revered as cricketer Sir Donald Bradman. In our household, these gentlemen were Australia's most famous people – they gained their knighthoods not because they helped to win the war or were a retired politician or the Mayor of Melbourne or Sydney. They were the world champions in their field of sport!

So, we strove to emulate Walter Lindrum on the billiard table, even though we had only a scant picture of how good he was or how he played, via newspaper articles and radio descriptions – we didn't have the benefit of close-up TV images! We might never have come close to his skill, but many a good hour was spent at the Billiard table.

The hidden secret of the piano was that it was actually a 'Pianola' – which meant it had more than just a keyboard which could be played in a conventional manner. Underneath the keyboard there was a fold out pair of pedals which when pumped by foot would activate a set of bellows, providing both compressed air and created a vacuum, which in turn both activated the elaborate mechanics which wound through paper rolls of music, and caused the piano keys to be automatically struck. The Paper Music Rolls, of which we had a cupboard full, had perforations punched in them, which when passed over a central horizontal metal unit, with a separate hole for every key on the keyboard, each with a tiny tube connecting back to a 'striker' would activate the striking of its key. So, as the roll passed over the bar, changing air pressure would cause each key to strike according to the musical score on the roll – the piano would play itself. It seemed almost magical.

This was an important part of our family entertainment in the days before television, and before such things as iPads and play-stations, which weren't even science fiction at the time. We spent many hours playing through our collection of musical rolls. And the pianola lent itself to family 'sing alongs', often when we had visitors. We would all gather round the Piano, and sing as the words of the song, which were printed on the roll as it passed by. I have great memories of this bygone time. I think it's a great pity that modern families (ours included) do not enjoy the act of singing to music as much as we did.

The Farm Ute

It seems to me that if you wanted one archetypical image of life on our farm, or any other farm in the district for that matter, you couldn't go past the Farm Ute. Or more specifically, a Ute with a dog in the back.

It is one of my great regrets that we have few if any good photos that show this core part of everyday farm life. Right through my childhood, and indeed through all my time on the farm, there was scarcely any activity that somehow didn't include the Ute. It was used for everything – even if it was just to drive from the house to the shed!

And the moment you got in the driver's seat, the dog (or dogs) jumped in the back. Once up in the back, the dog found himself variously sharing the space with grease guns, toolboxes, shovels, axes, drums of fuel, empty wheat bags, binder twine, rope, drag chains, hydraulic jacks and blocks of wood, iron posts and coils of wire, wire strainers, and more often than not, if they weren't at school, kids.

On this latter point, the issue of safety was not considered. It was perfectly usual for passengers to climb into the back. In fact, this was the main way we got to school. We couldn't all fit in the front, so it was into the back, along with our bikes and school bags.

The issue of safety wasn't just the risk of falling out. One of the attributes of a Ute tray is that it was nice and flat, made out of metal and or hard timber planks, nice and shiny when it was new. So that meant that all those items occupying the back were prone to relocating themselves, often rather quickly, depending on whether the driver took a sudden turn, or braked too hard. Take it from me, a hydraulic jack, a post-hole driver, or a jerry can of fuel, can make it from one end of the tray to the other in a hurry! This hazard was ameliorated in our various Holden Utes, whose trays were quite small and had the protection of a wheel arch which intruded into the tray. This was not so once the Toyota Hi Lux came along, with its quite large, flat checker-plate metal tray. Any dogs, or kids, in the back of that had to be very alert to the driver's intentions indeed!

Just about the best possible expedition in the Ute, certainly from the dogs' or the kids' perspective, was to go 'out around the sheep' The best part about going out around the sheep was that you were not constrained to driving along the road or one of the farm tracks – you had a whole 100 acre plus paddock to head off in any direction.

As such, there was no better place to learn to drive, from quite a young age. The main determinant was if you were big enough for your backside to be still on the front edge of the seat when you had the clutch depressed. Then, you were big (or old) enough to learn to drive. And out in the middle of a paddock, you had no obstacles to consider, leaving you free to concentrate on the use of the clutch, accelerator and brakes. It was the perfect place to learn, both to my benefit, and a generation later, for our kids as well.

Regardless of whether I was the driver or not, those trips out around the sheep were an important part of farm life. Observing how the sheep were doing generally, and looking for more specific husbandry issues such as fly strike was a regular task. And particularly through the summer months, to check the water troughs which were fed from the several water 'bores', or windmills and tanks, given there was no other surface water supply.

Often, the purpose of the trip out around the sheep was to move them from one paddock to another, which is of course where the dog got his piece of the action.

I have fond memories of many of our faithful sheep dogs, but one from the latter stage of my time on the farm stands out – the gangling, exuberant, sometimes empty headed, but always trying his best, and funny, 'Sniffy', a slightly less than pure-bred Kelpie.

Almost certainly one of Sniffy's main problems was that he didn't have the benefit of proper instruction in the art of being a sheep dog. I have to take a fair bit of responsibility for this. It took me another forty years, and the advent of a miniature poodle, to find if you had a reasonably intelligent dog it is amazing how you can train them to do things, building upon their natural instincts. This included a fair bit of time and patience, plus a pocket full of rewards or treats such as 'Smackos'. Now I'm not saying Sniffy was totally bereft of guidance and coaching, but it was mostly the yelling out of 'Go Right Back!" or "Come Here!" and then seeing what happened!

The first objective in moving the sheep was to get them all together in a mob, so that you could head them all towards the gate. So, off in the Ute in a big wide arc heading them into the middle of the paddock, and then it was Sniffy's moment. He would respond with glee to the call of "Go right back!", jumping out and hitting the ground about three metres away, roll over two or three times, pick himself up, and charge off around the left-hand side of the gathering mob, while we turned the Ute the other way, and headed back around the right flank, tooting the horn and revving the engine. Back on the other side of the paddock Sniffy was relying on barking for the same effect – all aimed at gathering in the mob. Just in case you haven't picked up on the apparent contradiction here, one of the issues arising from the inadequate rigour in Sniffy's training, was that whenever you called out 'Go Right Back!", he actually always went Left! Never to the right! To get him to go to the right, you had to yell "Come Here!" I guess it really didn't matter, as long as we understood which way we were all going to go.

Therefore, with Sniffy way out on one side of the paddock, and us in the Ute on the other, another one of his traits came to the fore. He would start looking back at us, wondering if we were going to issue a further 'Go right Back' instruction, and then take matters into his own hands (or paws). I think his thought process must have been 'It must be time to do a wheelie – that's what they always do in the Ute', so he would make a sudden change of direction, and come racing back towards us chasing his portion of the mob into the centre. The problem with this is that we would have about 250 sheep in a nice mob heading towards the gate, while there would be another 50 that had been cut off by his change of direction, all heading off back towards the far side of the paddock.

This led to some considerable admonition, and a further instruction to "Go Right Back, and get them all this time!", which he happily did, having first looked at us as if thinking 'Jeez, you are being pernickety this morning – I got most of them for you! Wasn't that enough?'

Years earlier, I recall a demonstration of a different way, with a different dog, to move a mob of sheep. This also involved a trip in the Ute, this time to go over to our neighbour Evan Evans place, for a bit of a catch up about the latest branch meeting of the Wheat and Woolgrowers Association, the cricket or footy, the weather, and anything else that came up.

One morning we turned up to find Evan there with his dog 'Toby' at the sheep yards, near the barn and shearing shed, which was set in the block of Mallee Scrub that also protected the house, machinery shed and other outbuildings on the Evans' farm.

One of the good features of those Holden (or Falcon) Utes was their nice smooth sided rear section which enclosed the tray, providing a perfect place to comfortably lean against to have a yarn for a half hour or so. This first involved both Dad and Evan getting out their respective tins of 'Havelock' fine cut tobacco, and small packet of 'Tally Ho' cigarette papers, to begin the task of rolling a smoke – spreading out a layer of tobacco on the thin piece of paper, rolling it up and licking the sticky edge section to seal it up, then poking each end in with a match head before then lighting up. I was born into an age when almost every man was a chain smoker.

With cigarettes lit, they settled in for their session, which I would be half listening in to while I bounced my footy in the back of the Ute. But not before Evan, having opened the gate into the sheep yards near where we had parked, turned to Toby, and said "Go right out back and get 'em Toby".

With that, Toby took quietly off, sauntering out along a clearing in the Mallee trees towards an open gate into an adjoining paddock, and then disappeared from view.

A good half hour, and two more cigarettes later, you could hear the sound of bleating sheep out behind the Mallee scrub, and then a whole mob appeared coming in through the gate and up towards us! Toby was at the rear, carefully supervising; making sure every last sheep came in through the gate. In five minutes, the sheep were all safely in the yard, Evan simply left with the task of shutting the gate. And giving Toby a very well-deserved pat on the head!

Toby had gone out across one paddock and into the next a half mile away, gone right around the paddock to make sure he had every last sheep gathered into a mob, then herded them back along a couple of fence lines, before bringing them in through the Mallee to the yard – all on his own, while we leant around the back of the Ute, waiting for him.

Toby was the best sheep dog I ever saw at work – I am sure he could have been a champion at any sheep dog trial in the country. And Evan must have been the best trainer of a sheep dog in the district.

School Days

The little wooden Schoolhouse at the bottom of the Perenna Hill – State School 4091, being the place where my (our) primary years of schooling took place, obviously has to be a part of this story.

This was, as were many others scattered across the landscape, a place of small enrolment. I believe that in my time as a student there, the maximum number of students in any given year was fifteen. It only served a few families. During my time, there were two Evans families represented (all cousins of ours via our grandmother, Elizabeth), plus the Morrisons, the Berlunds, and of course we Dodds'. That was it.

This necessarily meant that it was a single teacher school. That was, as previously noted, how Mum arrived in the district – as the single teacher at Hamlyn School, about 10 miles (15 kms) further south. One of my reflections of this is that the quality and ability of the sole teacher at the school was of great importance, as the entirety of your learning rested with this one individual, who had to be able to instruct a bunch of kids across all grades all at the same time. A rather daunting task as I have since come to respect greatly.

I believe I have much to thank one particularly outstanding person, whom I was fortunate to have as my teacher, for I am pretty sure for four of my six years at Perenna, the recently deceased Ron Attrill.

Ron was an emigrant from England, whose path to Perenna took him via a period as an enlisted soldier serving in Korea, returning to labouring work before enrolling as a mature age student, and training to be a teacher. I'm pretty sure Perenna was his first posting, when he was in his late twenties. I remember him as quiet, calm but firm, a definitely intellectual fellow. He was a stickler for detail, and while my primary schooling is all rather distant at over sixty years ago, one clear recollection of his instruction was his disappointment in my writing skills, making me undertake a period of revision before I moved on. That memory stuck with me – being annoyed to be told my work wasn't good enough and determined to prove him wrong!

In amongst teaching six separate grades, he made time to introduce us to the world of music. This is the aspect of my primary schooling I think I remember most. Not that our kids were particularly thrilled years later whenever I reached for my recorder and proceeded to play 'Greensleeves' or 'Amazing Grace' for them. Notwithstanding my rather un-stellar musical career, I look back on Ron's instruction in how to read and appreciate music, and learn the basics of playing a musical instrument, with a lot of gratitude. He was a good teacher, and role model. His influence was such that I don't accurately recall the names of any other teachers I had at Perenna.

Figure 9 Perenna School - (Editor's note) I think this is taken in 1957.
Back row from left is Des, then me. Shirley is in centre row, just left of me.

How we got to school wasn't much different from today. We were mostly driven there by our parents or rode our bikes. Correction, mostly driven by our fathers, as women didn't drive back then. Except, that is, for the Berlund girls, as we referred to them, five sisters who lived out on the edge of the scrub, near the Vermin Proof fence. They did drive the five miles and back to school every day – in their horse-drawn buggy.

How our teacher (Ron Attrill and those before and after him) got to school every day was also an issue. Not so much as to his method of transport (always a car) but where from? There was no township at Perenna with houses for rent or purchase, so where did he live? The answer was that all the local families were rostered by the School Committee to 'board the teacher'. This was generally on a term-by-term arrangement, as he rotated from family to family.

There are a few other aspects of my school days at Perenna to relate.

The first was the Annual Lake Hindmarsh School Picnic and Sports Day. This brought together the kids from four local small schools – Perenna, Lorquon, South Lake Hindmarsh, and North Lake Hindmarsh, at a nice flat open area used as a picnic ground by locals on the Western Shore of the Lake.

The lake itself probably deserves a mention here. One of the first things we proudly learnt about in geography and studying maps, was that this big lake that we lived by, was the largest natural freshwater lake in Victoria. Fed by the

Wimmera River, it is the first in a series of 'terminal' lakes, as the river does not run to the sea. Lake Hindmarsh is, or was, a beautiful big expanse of water when full, stretching some 20 kms south to north, and 8 kms across at its widest point.

It has been to my good fortune, that the Lake, during my 54 years of living beside it, was full or mostly full for most of that time. Records show it was dry in the late 1940s, following a series of dry years, and thanks also to significant diversion of water from the river as the Wimmera Mallee open earth channel system and associated irrigation schemes reached maximum development. It filled in 1950 and remained so until it dried out following the severe drought year of 1967. It filled again in 1973, and remained so until 2000, the year Mary and I left there.

Sadly, apart from a brief time when about a third of the lakebed was covered after a flood flow in the summer of 2010-11, and again in 2016, the Lake has remained mostly dry ever since. It only reached about half full in the recent La-Nina wet period of 2022. This is despite the completion of The Wimmera Mallee Water Pipeline and decommissioning of the open earth channel and irrigation schemes in the early 2000s.

If anyone still questions the reality and impact of climate change, they need look no further than Lake Hindmarsh. I recall reading a CSIRO report in the early part of this new century where a projection of the impact of climate change would be 20% less 'yield' or run-off in the river's catchment by 2030. Unfortunately, after 20 years, that reduction of flow appears to have been surpassed already.

Back to the Picnic, as the waves gently lapped the shore of our near full lake.

The many events for us kids included the standard fun activities of the day: egg and spoon races, (you had to balance an egg on a teaspoon held in your mouth and not drop its as you ran), sack races (you had to run or shuffle in a wheat bag), a legs-tied-together tandem race, and a potato race. The grown-ups were also catered for with a Blindfolded Wheelbarrow race for the Women, and a blindfolded Hammer-in-a-Nail event for the men. All good fun.

But the highlight, held just before everyone hopped into the piles of sandwiches, sausage rolls and cakes, washed down with cordial or tea, was the rooster chase!

Now this might be a bit un-PC (not politically correct) in today's more enlightened times, but it was considered a legitimate fun activity at the time. It relied on someone choosing the largest and most crotchety rooster they could find in their chook yard and bringing him along in a wheat bag. It didn't help the rooster's mood to be lying there beside a post all afternoon with his head stuck out of the bag he was in, so when the time came for him to be released, he was out of here!

The idea was the rooster was given about ten seconds head start, before the kids were allowed to set off in pursuit. Whoever caught him ended up taking him home – presumably to end up as a Sunday Roast. So, it was about fifty kids and a squawking, wing flapping rooster going in all directions down by the water, around the red gum trees, underneath the cars, up the Lake bank, and down again, until finally a victor emerged. I recall that this was almost always Joy Berlund! Joy won most of the races she entered, and was pretty fearless, definitely the gamest tree-climber at play time at school. She wasn't too concerned about a close encounter with an angry rooster.

Another day in the school calendar was the Annual Nhill Hospital Country Schools Sports Day, held at Davis Park - The Footy Ground - in Nhill. This involved all of the small rural schools in the district, so added up to hundreds of kids and their families.

This was a bit more formal than the Lake Picnic. These were real footraces, run in lanes marked out with white lines on the lush green turf of the oval. This was an experience in itself for us country kids, where if we were lucky there might have been a few square yards of back lawn at home. But first we all were assembled in our groups, behind our School Banner mounted on two poles carried by two of the biggest kids. Led by the Nhill Brass Band, we then marched onto the oval, and around the perimeter before being directed to our place out beside the racetrack. It was just like the Olympic games!

Now this marching practice was a good lead up to the next really big event of the early 1950s. The Royal Tour, the first time Queen Elizabeth II came to check us colonials out. This was a really big deal, for which we were primed months in advance. Having someone as revered as Her Majesty here to see us was seriously important! And I must add, was a fairly significant boost to our study of British, and therefore Australian History.

Of course, with all the swanking around in Sydney, Melbourne and Canberra, there was no time to come to the Wimmera, let alone Nhill. The only option was to see her in Hamilton – presumably chosen as a country centre for the Royal Tour due to the wealth and political clout of Western District graziers.

How we got to Hamilton was to me the really good bit. A very large contingent of Wimmera kids were all transported there by a special train. I well remember this – boarding the train at Nhill. It was hauled by two big stream locomotives in front of many old red passenger carriages. Off we went to pick up more kids at Dimboola, then more at Horsham, Murtoa, Stawell and Ararat before we turned south for Hamilton, with a train full of loyal young subjects, carefully watched over by many schoolteachers and I think a selection of adult dignitaries. I think it was my first ever train ride. What excitement!

Finally, we reached Hamilton, where we were all marshalled together to march up to the main street, Grey Street, which was closed off for the day, and then turned right and headed out onto the local footy ground, Melville Oval. Here's where that marching around Davis Park came into its own, as we made our way out to be assembled behind the many lines of rope all set out in long narrow corridors. And there we waited. Until along came a convoy of several Landrovers, with the most important one having a platform in its back tray. And there, standing and waving as she drove past, with of course the Duke of Edinburgh by her side, was the Queen. We all waved our Union Jacks like crazy, as she drove back and forth along the oval, between all the lines of school kids dutifully staying behind the ropes. And then she was gone. That was it!

I can recall a similar time about sixty years later when we stood in the street in Goolwa south of Adelaide to see the 'Tour Down Under' road cycling race. Standing behind roped off lines beside the road with hundreds of others, waiting for the cyclists to arrive. Then an entourage of cars came tearing around the corner, followed by all the cyclists who whooshed past all in about a minute, and they were gone. We stood there and looked at each other and said "Well, that was good!" "Now what do we do?" 'Go Home?"

Well, that was what we did that day at Hamilton. After five minutes of waving our Union Jacks, we marched back off the oval, back up the main street, onto to the train, and set off for home, many hours away. But we had been part of Australia's history as we saw the Queen. And actually, it really was a great adventure and day out, a special occasion. I'm glad to be able to say I was there that day, and I enjoyed my first train trip.

The final bit about the old school site at Perenna, was its co-location with the district's first Methodist church.

My main memories of this place of worship were that of a rather dark and foreboding place– its outside walls were unpainted, dark brown earthen 'Mud Brick', with a few small opaque windows, with a tin roof which spanned out to form a veranda all round – a core part of the design of a mudbrick structure to protect the earth walls from rain.

It was here that local farmer Mr Max Sluggett, a volunteer 'lay preacher', and (top-level member of the Lorquon Cricket Club!) used to gather up us kids and take us out the front door of the church around to a lean-to on the side veranda for our Sunday school session of Religious Instruction, before re-joining our parents after the sermon by the visiting Methodist Minister of the day. Max spent I think probably a couple of years by simply reading to us excerpts from the Old Testament – several verses at a time, each Sunday. Going to church, having the unquestionable edicts of the Bible inculcated into our upbringing was a fundamental aspect of our childhood and early education.

One of my standout memories of going to church at Perenna was of the annual Harvest Thanksgiving Service, where the church was decorated with sheaves of hay, bushels of wheat and oats, jars of preserves, and flowers, fruit and vegetables, for which we all gave thanks, and considered the plight of others not so fortunate.

Pertinent to my later focus on better environmental management, I can remember going home in the car from one such service with Mum commenting on how lovely the Bridal Veil Creeper that Frieda Morrison had brought along to decorate the pulpit looked. Bridal Veil creeper (Asparagus asparagoides, an import from southern Africa) is now one of the worst environmental weeds in Victoria, with large sums of money granted by government each year to assist in its control.

Like so many other school sites, the little square of land on which both the church and school stood has now reverted to private ownership, subsumed back into the adjacent farm. If you didn't know those institutions which held such importance to us at the time did exist, then you would drive straight past without looking.

Down the Well

As with the settlement of pretty much all of inland Australia, the availability of water was a fundamental requisite for our farm. While in wet years there was quite a bit of local run-off and pooling in low areas, there were few, if any permanent sources of reliable surface water in the Nhill district, necessitating the scooping out of dams to a depth that allowed water to be caught during winter and stored through the hot dry summer. There really weren't any such dams on our farm, but what was available was a source of underground water. That was good, but the problem to be overcome is that it was about one hundred feet down - thirty metres.

This was accessed by digging wells – vertical shafts, lined with timber, dug down to the water table, in fact several feet below it, to ensure a good flow available to be pumped. Pipes were lowered down, attached to the wall, beside an access ladder. A windmill was erected above to provide the lift needed to activate the pump down below.

So pretty much every farm in our district had a well, no doubt one of the first steps in establishing a new holding. I don't know who really dug and lined the well that we had at Pine Flat, or when, but I imagine it would have been people who either themselves or their fathers before them had learnt the skills on the goldfields. Some of the stories I recall spoke of teams of Chinese labourers being well sinkers as settlement progressed.

The well at our place was no longer used when I was young, having been replaced by the newer technology of a bore. One of the reasons for this was that the wells had been sunk to what was called 'the first water'. This first aquifer was quite brackish or salty – okay for livestock in the absence of any other source, but not ideal. Somewhat deeper down, was another aquifer, or 'the second water' which was less mineralised, and really quite suitable for most stock and domestic use. This aquifer has provided a reliable water supply for most of the West Wimmera since settlement.

To access this water, bores were drilled, and lined with a heavy, mostly six-inch diameter, steel pipe. These bores were drilled down to about two hundred feet into this 'second aquifer' Once accessed, this water would rise under pressure up to and beyond the level of the first water – at our place it came up to about seventy feet or twenty metres. Another narrower pipe with a pump attached was then lowered down the bore to just below this water level, with a windmill above to operate the pump and lift the water up to a corrugated iron tank sitting atop a big timber tank-stand. These were a familiar sight across the district, and still are to an extent, but many have been replaced by more efficient electric pumps, increasingly now solar powered.

I never got the chance, or was ever brave enough, to climb down the well. I can clearly remember it, covered over by a couple of sheets of corrugated iron with a block of wood on top. Safety first! Occasionally, Dad would remove the cover, and let us gaze down in wonder, and a degree of trepidation, while he recanted his favourite story of the day he had to climb down to try and fix the pump and found halfway down a rather disgruntled snake that had fallen down and was caught on a ledge!

So, what happened to the well after it became obsolete, and in fact an accepted danger or hazard for us kids? We filled it up with rubbish! This was accelerated by the advent of the Bedford Truck, with its 'GWell' underbody hoist. You could load it up with any rubbish or objects that were considered worthless and no longer of any use, back the truck up over the top of the well, operate the hoist and watch the load disappear down the well, clattering its way to the bottom.

That was probably the fate, or end use, of all the wells throughout the district. Just as well it was pre–Environmental Protection Agency (EPA) days, or maybe it's best kept quiet, and no one goes looking at what might be leaching away down there in the water table! At least it was prior to the large-scale use of agricultural chemicals, so I'm fairly certain that no empty 24D ester drums (24D Ester is a rather volatile herbicide) or other such hazardous items found their way into the bottom of the well.

But I do know about a lot of stuff that went down there that was rather a pity! Like the old 'mangle' that used to sit on the side veranda for flattening out sheets and towels and the like – considered obsolete following the purchase of a new thirty-two-volt electric iron! The mangle was a beautiful old piece. Down the well! And all of the gas light fittings from the house – also of no use to anyone following the advent of electric lights – down they went. There were lots of things went down that well that would be considered curios of some interest now.

Set against how we now handle waste disposal – and increasingly recycling, the disposal of rubbish down the well might seem rather questionable, but at the time it solved another problem. That was the dumping of rubbish in the open. Rubbish disposal up until probably at least the 1970s was just something you sorted out yourself, with no particular regulation by, or concern from, authorities.

Just about every one of those small patches of Mallee scrub, or timber paddocks on farms right through the district had its own rubbish tip – just an area slightly out of sight, where everything was dumped – from tin cans and glass jars and bottles, through to old cars and machinery. All left to rust and decay in the open.

Thus, it was quite an advance in tidiness to tip it all in a hole in the ground –such as a disused well.

Windmills

Figure 10 Our 'Southern Cross' Windmill

You couldn't be raised on a farm in the Nhill district and not have a strong recollection of windmills, pumping up all of that bore water on which the farm depended. Certainly, they looked impressive, and have been the subject of many a photographer, professional and amateur. Also, how they sounded. The sound of the windmill became very familiar, the rush of wind through its fan blades, the squeaking of its bearings as it turned, and the slow and steady sliding of the pump rods up and down, and the water flowing into the tank. There was a bore and windmill quite close to our house, and it was a steady, reassuring sound as you lay in bed at night, listening to the windmill outside.

The other great thing about windmills, is they were mounted on top of a metal stand – some ten metres high or more. The mill at home was quite high, with a skinny metal ladder up one side, so that you could clamber up to service the machinery at the top. That was for the men about the farm – Dad's job! For kids, it was a great place to climb up to see the view. You could see down on the roof of the house, look at the water in the tank on the tank stand, and look out over the trees to the paddocks beyond. Until Mum or Dad saw you and ordered you down straight away!

Another thing about windmills was that from time to time something went wrong, and needed to be fixed, or serviced such as topping up oil in the gear case. An OH&S nightmare!

This meant firstly pulling down on a long lever located at the bottom of the tower, which was connected by a length of wire and chain up through the turn table - the windmill was able to swivel around 360 degrees depending on wind direction - to the back end of the windmill's 'tail'. Pulling this down and securing it drew the tail around more or less parallel with the fan, the effect being to turn the mill around away from the wind direction and slowing it down and hopefully stopping. Unless you were trying to do it on a really windy day, in which case it would still keep turning.

Then you would climb up the stand to a wooden platform located just below the fan. Then reach up and feed the piece of chain, which you had draped around your neck while you climbed, around the tail strut and a fan blade, and securing it. This would prevent the mill from spinning (a rather good idea) while you then proceeded to do what had to be done. This would entail climbing higher up again, so that you could locate the bolts that secured a metal cover over the gears and workings of the mill. That would enable you to check the oil level and top it up if required or check if there was anything worn out or broken and in need of repair.

If it needed fixing for whatever reason, it was time to call in the experts! Enter a couple of real local characters, Mick Stevens and Keith Klemm.

Mick and Keith were a couple of locals who knew everything you needed to know about bores and windmills, and then some. Farmers themselves, they knew the importance of a continued water supply, so would pretty soon turn up in their old 'buckboard' truck which was something to behold. It had racks at the front and back which carried several twenty-foot sections of 2 ½ inch pipe, more similar lengths of pump rods all mounted above the roof. In the tray were boxes and boxes of pipe fittings, pipe wrenches and plumbing tools, windmill cogs, and spare pumps and pump parts. There was a tow hook at both the front and back, and hanging wherever it could fit, a set of pulleys and a couple of rolls of cable. And most importantly to me, they always seemed to bring along in that old buckboard a good sense of humour. It seemed they enjoyed fixing up bores and having a yarn with their customers while they worked.

For this reason, we would jump in our Ute, and escort them to the windmill in need of attention and hang around to watch them work.

The most likely reason why the mill had stopped pumping water was not that something in the works up top was broken, it was more likely that the pump 'washers' down below were worn out. This would require Mick and Keith to 'pull up the bore', to find and fix the pump, and then lower it all down again. This would entail setting up a pulley near the top of the windmill tower, and feeding a cable up through this, and connecting it to a special clamp apparatus that grabbed onto the top of the vertical bore pipe and pump rods, which were first disconnected from the windmill gear box. The bore pipe was actually a series of four or five twenty-foot sections (Sorry, its best to stay imperial here!) all screwed together, reaching the eighty foot or so down to the water level in the bore. Together, this all added up to quite a bit of weight, certainly not able to be hauled up manually. It was (is) all really a mini version of an oil rig.

The loose end of the cable was attached to the tow hook at front of the buckboard, and with Keith operating the clamp and process from a perch halfway up to the top of the windmill, Mick would jump in the driver's seat, put her in reverse and back up, pulling the bore pipes up from below. As each twenty-foot pipe came up it was unscrewed, the clamp assembled on the next section, and the old truck backed up again, until the final section, with pump screwed to the bottom was brought up. It was then installed with new leather washers, ready for the process to be reversed, and the pipes lowered back down.

The part I most remember about watching Mick and Keith work was how they managed their tools. Undoing the pump rods, and unscrewing the pipe sections was Mick's job, perched as he was halfway up the windmill, probably twenty feet up. Some quite heavy-duty plumber's tools were required. Thus, Keith would stand down below and wait for Mick to call out when he couldn't get a tight joint undone: 'Hey Keith – I need that eighteen-inch Stillson wrench – could you chuck it up?' Keith would ruffle around in a toolbox, locate the wrench, which probably weighed a couple of kgs., and come back to the base of the stand. "Here it comes' he would yell, and toss it up to Mick. He was a beautiful throw – Mick would just reach out and grab it as it hovered beside him! A minute later he would yell "I need a bigger hammer – could you chuck one of those up?" Back to the back of the truck would go Keith, have a look around and yell out "What about this one?" "Yeah, that'll do!" And up would come a blacksmith's hammer – to be easily caught in one hand by Mick as he held onto his perch with the other. They hardly ever missed or dropped one. Just as well, because hard hats were not a feature of this worksite. We reckoned Keith could make a heavy tool hover beside Mick for four or five seconds till he needed it!

They were good operators and knew what they were doing. It was and is an exacting task. If you make a mistake with a bore half pulled up, and it drops through the clamp, it's gone for good, down to the bottom of the bore, with the risk that another new set of piping and pump might not be able to be lowered down to the water. The bore would then be no longer useable.

Itinerant Visitors

It was a far from lonely existence out at Perenna – just about every square mile equalled a farm, with the family home, and often a worker's hut. Lots of people, with lots of interaction, not just for church on Sunday or taking the kids to school.

And this was regularly broken by a series of itinerant farm visitors.

Stories of Indian Hawkers moving about the country with elaborate horse drawn carts loaded with exotic items for sale are legendary, and apparently there was one who had a run through the district, but not in my memory.

One visitor I do have recollection of was 'Rawleigh Jack'. I recall him as quite a character, who quite regularly called on us, dragging a big suitcase out of his old car, and (at Mum's invitation of course) brought it in and onto the kitchen table. The case was full of Rawleigh's products – as I recall a mixture of ointments and medicines, plus spices and essences. He must have sold lots of Rawleigh's ointment to us, as I can remember just about every time we lost some skin with some sort of scratch or graze, out would come the round tin, and some rather smelly ointment was applied. I think he generally managed to add to our supply of vanilla essence, plus another packet of tapioca pudding. I always hoped it would be the chocolate tapioca! Which reminds me – it's probably sixty years since my last tapioca pudding – I'll have to see if I can buy some!

Rawleigh Jack was one of those real Nhill characters. He apparently had quite a history as a farmer, a miner and a trader of merchandise – all while raising a family of fifteen with his wife – thirteen of their own and two adopted4.

Bob Lineker, a tailor from Ballarat was also another trader whose visits were somewhat less often than Rawleigh Jack. His visits were always well anticipated, and involved staying at least overnight, sometimes longer. He did become a family friend and was the maker of much of our clothing – particularly our 'going out' clothes. Bob was actually the other passenger in the car that Jack was driving when they were both fatally injured coming back from the Trots at Moonee Valley to Ballarat.

Bob, and his son Lin, who was his apprentice, would stay with us until we were all measured up for new suits, coats, trousers and dresses for the year ahead. Then they would be off, to make similar visits to other regular clients, before heading back to Ballarat to make all of the clothing and complete their orders. Some weeks later the Mailman would turn up and deliver a pile of packages, neatly wrapped in striped, brown paper. Our new clothing had arrived.

It was one of the social norms that when you went out anywhere, not just to Church on Sunday, that you were well dressed. The casual age of denim jeans and cotton tee

---

4 Karen Reichelt-Cramer, Treasures of Nhill and Districts (Facebook Page) May 2020

shirts was a long way off – in fact the very suggestion you might consider wearing anything resembling denim was met with scorn by Mum – you wouldn't go to Nhill in 'working clothes"!

Accordingly, the regular visits by the Linekers were an important part of life, ensuring we did meet the required standard of the day.

Another regular visitor to the farm was the Methodist minister of the day, who did his rounds of the flock, devout or otherwise, which worked well for him at Perenna because most of the families there were Methodist. I recall his visits being announced by much barking of dogs, with Mum checking through the curtains to see who it was, and on saying, with a slight tone of unease – "Oh, it's the Parson!", she would quickly tidy up the kitchen and comb her hair, while he managed to settle the dogs down and come in through the gate to the back door of the house to cheerfully announce himself. I don't want to pass this off with too much cynicism – I believe that regular visits such as this were one of the really good and beneficial aspects of the much more church going life we had back then, depending of course on the personality and qualities of the minister of the day - not the overbearing chap who banned dancing at MYF!

Other occasional visitors to the farm were blokes trying to sell stuff. A particular variety of these were welcomed to most local places they called in to because they were good footballers! Always seeking to build on the local talent they had, the Nhill Football club from time to time were alerted to the possibility of attracting a noted football talent, be it as coach or just a good player, to move to Nhill and join the team. Understandably, this generally depended on being able to offer a suitable job to attract someone to make the move, on top of the match payments that could be made. One of the possibilities was to have them become representatives of some company that may wish to increase its profile and sales in the town and district, such as selling life or general insurance, or some other product or service.

One of these sporting imports turned up one day and found Fafa up for a talk. So, of course, the discussion for the first half hour was all about football. How good Nhill's side was this year, and how they were a real premiership chance. But the very slight change of topic which greatly increased the chance of a sale was when he agreed with Fafa that Collingwood were absolutely crucified by the umpire during last week's loss to Carlton. They would have won if it hadn't been for such blatant biased umpiring! By this time, I think Dad had turned up, and they were in complete agreement about this umpiring travesty. You could hear that the crowd was up in arms on 3DB's[5] radio broadcast, and the match report in the Sporting Globe backed up how many more free kicks Carlton had got! Of course, the umpire was biased!

This was the perfect prelude to asking Dad and Fafa about their supplies of farm lubricants, the end result being that they agreed they needed two forty-four-gallon drums of grease. Now I can remember thinking that I knew we had lots of farm

5 3DB was a commercial radio station in Melbourne, which was relayed to the Wimmera via 3LK, a local station with its transmitter on a tower at Lubeck, between Stawell and Murtoa

machinery with grease nipples that needed the occasional shot from a grease gun, but eighty-eight gallons (four hundred litres) was an awful lot of grease!

The two drums were duly delivered the next week, and I know that there was still a lot of grease left in a drum that was in the corner of the machinery shed twenty-five years later! It was pretty much solidified, and I think we had decided that it was perhaps better to be using a small tin of more refined modern grease on our new machinery.

Another bloke who must have been a smooth-talking salesman also did a successful tour of the district. I don't think he had anything to do with football, but he did have lots of forty-four-gallon drums on the back of his truck, all full of silver paint. Plus, a pump, coils of hose with a spray nozzle, and a couple of ladders. His target was all the corrugated iron buildings in the district, of which there were many. His service offering was to give all of the sheds, barns, and of course houses, a shiny new coat of silver paint, assuring his potential clients that this would ensure the galvanising would last for years more before rust set in if he sprayed his paint on – quickly and without any fuss.

He must have been efficient, both in his sales pitch, and in starting up his pump and spraying silver paint, because one morning we went off to school and I don't think there had ever been any discussion about painting any buildings. When we came home, there in all its shiny new glory was the barn – completely silver, including its wooden doors that used to be white, and a fair bit of the surrounding grass as well, and the roof of the house – reflecting the afternoon sunshine like a mirror! In the space of a day this bloke had turned up, convinced either Dad or Fafa of the need to do some painting, completed the job, and moved on to the next farm and customers. Within a week or two there were shiny new sheds and house roofs all across the district.

It's just as well we didn't know anything about lead poisoning at the time, otherwise we might have been a bit worried about the rainwater that ran off into our tanks to become our drinking and general house use water in the following months. Maybe we escaped any serious side effects due to a lot of the sliver paint flaking off and blowing away in the subsequent summer heat!

There were of course another class of travelling salesmen who toured the district – encyclopaedia salesmen. Now maybe encyclopaedia salesmen have received a bit of a bad rap over time, but I believe their wares were of somewhat greater benefit than gallons of grease and doubtful quality silver paint.

Along with many of the other books, magazines and newspapers we had to read in the pre-television era, there was always a set of encyclopaedias taking pride of place on a shelf of their own, and thanks to the sales ability of the respective representatives, these were from time to time updated, as they needed to be, given the explosion of knowledge and technology through the fifties, sixties, and seventies. The earliest set of such reference works are still in our possession, and date from 1935. I believe this set – the "Modern World Encyclopaedia" is the set that Mum said she put her first pay check as a young teacher towards.

Fridays in Nhill

Friday in Nhill was a part of the social scene of the farming community that surrounded the town. Nhill, of course, was the major service centre for the district, certainly through the 1950s, 60s and 70s. Thereafter, more and more services, such as government departments, machinery and motor dealerships, and others steadily withdrew (or were withdrawn), with Horsham taking over as the main service centre.

I suspect that the social institution of Fridays in Nhill was at its peak in the immediate post war years. Everyone had a car, roads were steadily improving - a trip into town wasn't the undertaking it used to be. But the main day when most people went into town was for Friday afternoon. It was a day of business, shopping and socialising.

I recall that this meant that most of our school lessons at Perenna state school would need to be over by lunch time on Friday, as most of the kids would be picked up early to join in the trip to town.

The first thing about going to Nhill on Friday, was that everyone was well dressed. The men would almost all have a coat, tie and hat, while the women would wear their best outfit, including fancy hats and gloves. Many of the younger men would be proudly wearing their football club jackets, one of the many lines of apparel, men's and women's, available at Nhill Stores. No trip to Nhill could not include a visit to Nhill Stores – the town's major emporium, at its prime location in Victoria Street. It was the equivalent of going to Myer in Melbourne!

Nhill Stores catered for a wide range of needs. It had a grocery department, ladies and men's wear, and haberdashery departments, as well as home furnishing. Out the back was the hardware department and timber yard. The indoor departments were set up across the ground floor, with the manager's office and cashier department on a mezzanine level at the back of the main building. The place buzzed with staff, most departments having a manager and at least one assistant.

And the place was at the forefront of modern technology, right up there with today's 'tap & pay' transactions! As a kid I was mesmerised by the amazing system of wire cables stretching from the main counter of each department up to the cashier's office. Each cable had a spring-loaded lever at its end, which catapulted a little brass canister with little wheels attached to the cable, which held a customer's purchase docket and tendered cash payment (say a 10 pound or 10 quid note), up the line to the manager's office. The cashier, sitting up aloft, suitably dressed in a vest and wearing an important looking hat, would unload the canister, process the transaction, and the brass canister would come zinging back down to the counter with the receipt and change! How high tech was that?

I don't know when the Nhill Stores system of cables and brass dolly's and canisters was dismantled, maybe it was at the time of conversion to decimal currency - 1966. The last time I saw such a system in use was in a store in Charters Towers in North Queensland in 1981. I can remember buying a pair of socks I didn't need just to re-live the memory of Nhill Stores.

But there was much more to Nhill than Nhill Stores.

The first thing to note is how regulated business hours were, with the Town Siren at the core of the rigidly followed trading hours. I reckon the siren must have been a leftover from the war years. It was mounted high up on a pole at the front of the power station, which was located just northeast of the town's still standing and very impressive Post Office building. It was loud! You could hear it from miles away. It was clearly audible from up at the high school where I spent my latter years of education, quite a way from the town centre. And it went off without fail at 8.30 am every morning, (Monday to Friday) to denote the time for shops and businesses to open their doors, which they duly did. It next sounded a 12.30pm, which denoted lunch hour. As the blaring sound faded, all businesses (except the pubs and cafes) ceased work and closed their doors. The owners and workers would variously get in their cars, hop on their bikes, or walk off home for their midday meal. However, they made sure they were back on time for the siren to go off again at 1.30pm, opening their doors again for afternoon trading to begin. Commerce then proceeded until the final siren blast at 5.30pm when business for the day promptly ceased, and everyone went home. Well, not everyone, because those (males) with the inclination had a half hour left for a few quick drinks at one of the town's five pubs before their closing time a 6.00pm - the famous six o'clock swill. No siren to advise of that time, which maybe was not quite as rigidly held to, unless the local constabulary chose to make an appearance, which I understand was generally with adequate forewarning.

I never did find out whose job it was to press the button to set off the siren. I'm sure it was well before any automated process, but whoever it was, they were always there at the correct time. It was an interesting insight into how regulated, at least in some ways, society was back then, compared to today.

Where else did we go on our Friday afternoon in Nhill?

A weekly stock up on groceries was always a priority. There was more than one grocery store in town, but our preference was always Whiteheads Store, on the opposite side of the main street to Nhill Stores. Whiteheads was, like Nhill stores, divided into "departments." The Grocery Department on one side as you entered, with household needs and hardware opposite, while up the back of the store was, for me, the much more interesting department stocked to the bleachers with sporting goods, toys and comic books!

So while Mum stood at the grocery counter with her list, the storemen, very neatly attired with collars and ties beneath their white shopkeepers coats, scurried off to their wall full of shelves, often requiring a ladder to reach the required product up near the ceiling, as well as under the counters, to retrieve a box of this, a packet of that, two pounds (weight) of that, as she read out her requirements to them. It all built up to a big stack of brown paper bags stacked neatly on the bench while the manager entered each article and its price (in Pounds, Shillings and Pence) into a docket book, to eventually produce a handwritten invoice. The self-service supermarket was still a couple of decades away!

The grocery shopping was a necessarily long process in time, a perfect opportunity for kids to head up the back. How I wish I had a photo of this Aladdin's Cave of stuff for kids! It was a joy to behold. It was our good fortune that Mum and Dad both saw the value in us having the benefit of many of the products available over time. Of course, I went straight past the dolls section, which was of more interest to sister Shirley. I couldn't wait to see the latest 'Meccano' Sets of which we built up over time a pile of pieces at home – a bit like todays' kids' Lego. Meccano was my favourite toy among many. It was made up of lots of metal parts – plus small wheels, gears, axles and pulleys. Along with dozens of little bolts and nuts, with a screwdriver and spanner set to suit. As in Lego, the pieces were boxed to make a specific model if you followed the directions. Or you could freelance and try to make any structure or machine you could think of.

But there was more in this wondrous place. We gradually built up a big collection of 'Bilda Bricks', a predecessor of Lego that fell apart very easily! And lots and lots of 'Dinky' toys (small metal models) – we built a collection of little cars and trucks to push around in the dirt out at Perenna. There were cricket bats, tennis rackets and balls, footballs and basketballs of course.

And then there was the hobby and board games section to browse, plus shelves of books to read. I am bit hazy on this, but much of my youthful reading was the 'Biggles' series, and Enid Blyton Books, and I think most of those were purchased in Whiteheads. And given that stamp collecting was a popular pursuit, there was a whole counter full of packets of used stamps from all over the world.

And the comic book rack! This was the glory days of these publications – most notably the "Walt Disney' series. These were my favourites – in particular those that featured Donald Duck, his nephews Huey, Dewey and Louie, and Uncle Scrooge McDuck. And his nemesis the 'Beagle Boys'. We built up a library of perhaps hundreds of these. I have no idea what happened to them as we grew older and lost interest – probably went down the well or were used a fire -starters. Seeing them now in museum collections makes me rather regretful about their demise.

And leaving the best part of the back of Whiteheads store till last, there was the Fireworks Section! I know there are many good reasons why fireworks are now banned and unavailable for purchase by the general public, which makes it all the more amazing that as a twelve or fourteen year old with a ten shilling note to spend, I could walk out of that store with a box full of Threepenny Bungers, Penny Bungers, Sky Rockets, Roman Candles, and braces of 'Tom Thumb' crackers.

Now a threepenny bunger was a serious cracker! I can tell you from personal experience. They were about three inches (8cm) long, and if you lit the wick, (right on its end, not in the middle!) then carefully but fairly quickly placed an empty Jam Tin over it and stepped back, there would soon be a massive 'bang', and the tin would be catapulted many metres skyward, before clanging back to earth (or onto the roof of the house) with a very rounded out bottom end. Maybe my observations thus made of soaring jam tins and skyrockets gave me a very good understanding of the level of risk and excitement attached to the early launches of astronauts into space, as the space race of the 1960s and subsequent moon landings unfolded.

I know it was fun, and I avoided any serious mishaps, but I am very glad that you could no longer purchase these things by the time our kids reached that age! The level of risk was extraordinary to kids, and big kids – adults - alike. There were two times of the year when Whiteheads were well stocked with Fireworks – prior to Guy Fawkes night on November 5 and 'Empire' or subsequently 'Commonwealth' Day in late May. And every time would yield stories of severe burns, injured eyes, even lost fingers as a result of these celebrations. That was one parental worry we didn't have!

By the time we had visited Nhill Stores and Whiteheads (Mum and us kids – Dad was off talking to anyone he could find about football, council and other business as he gravitated towards the bar at the Commercial Hotel) it was time for refreshment at the Central Café. This was our first choice among many such establishments. Right in the middle of the main street, its shop front was a very impressive stained-glass façade with an arched recess for the doorway. That façade is still there today. Once inside you found the milkbar counter on the right, and a fruitery section on the left. Then you moved into the seating section. This too was most impressive – with several varnished timber and plywood booths with padded seats and tables, with an individual light and mirror on the wall. The back wall had a double swinging doorway through which the waitresses could hurry back and forth from the kitchen.

We would find a booth to be seated, which was quite often well towards the back, as the place was frequently packed on a Friday afternoon and wait for our orders to be taken. It would be milkshakes for the kids and a pot of tea for the ladies plus a plate of freshly made sandwiches and cup-cakes. All very civilized for the time – but definitely not a cup of coffee in any form!

Now there's another subject – coffee! I am sure that well into my late teens I would never have seen a cup. The first offering was probably made using 'Bushells Coffee Essence' which you poured into a mug of hot milk. It was sweet, icky stuff that never really appealed. It was probably not until well into the sixties that us British Heritage tea drinkers began to be lured to coffee.

An early but I think a false start was made into this brave new world of coffee shops by local entrepreneurs Bongiorno Brothers, who already had their hands in several businesses in town, including a car dealership and garage, plus Nhill Aerated Waters, where all our soft drink came from. Their new business was the Everest Café, around in Nelson Street, and pride of place at the end of the main counter near the entrance was the first Espresso Coffee machine I ever saw. It did look large, complicated and presumably expensive.

Now maybe it was just us that never became immediate converts, but it never seemed to me to have much if any use. I can remember asking Dad when we first checked out this new shop "What is that thing, and what does it do?". To which his reply was something like "It's something the Italians (or the 'I-Ties') use to make coffee with, but it grinds and hisses and steams and takes so long to make, I'm buggered if I know why they bother!" "Only Italians drink it, I'd never touch the stuff!" My enjoyment of a well-made coffee was a decade or two away yet!

I've often wondered since how good a coffee this new machine produced, and how wide the custom was. I suspect not many were persuaded to join the new trend, as from my memory the Everest Café did not last for very long. Our British tea drinking heritage dominated our choices! The interesting thing is I think the Bongiorno Brothers (Sam, John, Bob & Vince) and their father before them probably had a longer Australian background and citizenship in Nhill than many of us Anglo Saxon families.

There were a few more shops to visit before our afternoon in Nhill was over. Just up from the Central Café, past the store of GJ Coles & Co, was Edwards Shoe Store. Its proprietor was a very neat, dapper well-spoken man, Eric Edwards, who ushered the family in, having correctly remembered all of us kids' names as part of his greeting. He was a born businessman and salesman! The walls were stacked to the bleachers with boxes and boxes of men's, women's and kids' shoes – the whole place certainly smelt of leather.

But pride of place in this store, in this post war period of new technology and innovation, was right up near the entrance – the foot X-Ray machine. I recall this was housed in a rather large wooden cabinet, probably four feet high (a bit over a metre – just too high for me to peer into the viewing slot on its top!)

The purpose of this device was to see if your chosen new shoes fitted perfectly. As a result, we would be fitted up with a pair of shoes, and directed up to the machine, where you stood on a little platform and placed your foot into a slot at its base. Eric would turn on the switch and peer into the view finder – and after a brief period of consideration, would say "I think that's a little tight, we might have to go up half a size", "Have a look in Margaret, and see what you think." Mum would have a look and confirm Eric's finding. I would be standing there wishing like hell I was tall enough to be able to peer down the view finder and see what my foot looked like myself! I was imagining it must look like the white and black bony structure of a foot I had seen in some science fiction comic book.

I don't know how long the X ray machine stayed there in Edwards Shoe store before the realisation that too much exposure to X Rays might not be a good idea.

I don't think I have ever suffered from over exposure – my feet and legs seem to be still OK. Although, maybe it does explain my propensity to fail to end my golf swing correctly with most of my weight on my left foot, with just my right toe touching the ground!

On the subject of technology, two really busy, and I imagine quite profitable businesses in town were the two electrical appliance stores: Bounds Radio, just beside the Central Café, and pretty much straight opposite, Shultz's Electrical. One or both was almost always on the agenda for a visit on a Friday afternoon, having now been joined by Dad who was interested in this stuff. They must have sold hundreds of new television sets after TV was introduced just in time for the Melbourne Olympic Games in 1956 – although television reception was not available until the early sixties as I recall in our country area. And then in turn, they would have sold hundreds more when colour television was introduced in the early 1970s.

As a consequence, it would be into Bounds for a discussion with its proprietor Frank Bound, who would have sold us our first 'HMV' TV set. The discussion would often be about how we could improve our reception – because the image we saw mostly out at Perenna in the early years of TV was very 'snowy' or fuzzy – barely viewable. This was due to the fact that we had to try to tune into the only transmitter available to us, on Lookout Hill between Ararat and Ballarat – a long way away.

To try and overcome this, Frank had already sold us a fifty-foot-high cable stayed antenna pole, and a much bigger antenna to mount on it. The next step was to move to a new high tech coaxial feeder cable, fitted with a 'booster' box. This would go into the boot of the Holden for the trip home, along with the groceries, comic books, shoes and whatever else we bought that day.

And not to be missed before it was time for home was Bullen Bros garage. Dad would announce that he had to go round there to pay a bill, and "did we want to come to see the new model Holden that had just arrived?" Did we ever! Thus, off around into Nelson Street past the Commercial Hotel, a couple of stock and station agents, the 'Spot Café, and past Fischers Motors – the Ford Dealers. But we never went in there, because we were Holden people! We wouldn't even go in if they had a new Zephyr or Falcon on display. We just didn't go there. It was a bit like why we never had seen the inside of the Lutheran Church – we weren't part of that flock! We were Holden People!

No, it was further up the street past the Picture Theatre, and into Bullens – a very impressive, expansive building with a front showroom, where four or five new Holdens could easily be displayed. It was always a highlight to sit in the latest model and enjoy the smell of its newness. And to discuss how the order for our next car was going.

Cars were certainly a focus, and in a manner repeated by the youth of later generations, driving your new or not so new car up and down the main street was a statement of some sort! I'm not sure exactly what, but anyway it was. And in Nhill's case made all the more impressive by the wide divided main street with one-way traffic each side. As cars did not have such advanced features as turning indicators in the early fifties, this meant you had your driver's window wound down, with one arm resting on the door sill (almost always with a lighted cigarette in hand) ready to stick your arm straight out to signal a right turn, or in a halt signal if you intended to slow down or stop.

One aspect of doing these laps up and down the main street was that it was all men. It was pretty much a statement of fact that women weren't expected to and didn't drive. This slowly broke down over time, but not quickly. I do remember the commentary that ensued as the odd brave woman broke this then social norm and was seen driving down the street.

The lack of equal opportunity for women was very real throughout my childhood and formative years, and the breaking down of those barriers was a slow, and still ongoing, process.

Our day out in Nhill was nearly done. The last bit would be to sit in the car from closing time at 5.30 (as alerted by the siren) while waiting for Dad to conclude his important council networking in the Pub before the six o'clock close. Then it would be around to the Newsagent's at "Buzz's Corner", where to overcome the strictures of the rigid trading hours, owner Max Bussau had approval to install a small side window opening where you could buy the Melbourne Herald newspaper afternoon edition, which arrived in the late afternoon every day. That was a must, as it had a big sports section where Saturday's round of VFL football was extensively forecast, or in summer the Test Cricket. The weekly 'Truth' newspaper was also another occasional Friday night purchase, but its importance, particularly the pictures of ladies on page three, did not become apparent to me until my teenage years.

The Road home from Nhill

One of my earliest memories of driving home on a Friday evening from Nhill in our Holden car, heading north from town on the Netherby/Lorquon Road must have been in the early 1950s. The two key things I recall of that trip were that the bitumen surface ended just at the end of the golf course on the edge of town; and looking across towards the Airport, which seemed like a fairyland with a whole array of red and white lights blinking on to mark out the various towers, water tanks, and radio masts. The airport was a key wartime base but continued as an important aspect of life and travel for the district really right up until the present day. Although it would never reach the time again when you could buy a seat on a regular passenger service to Melbourne, flying in a Douglas 'DC3', as I subsequently did in the early 1960s.

So as the airport lights twinkled away in the west, we began the often slippery twenty-mile trip on a gravel road, eventually petering out to a dirt road for the final three miles home. If this all sounds rather challenging, given the wet years of the 1950s, maybe it was. But it was what it was at the time, and we would have accepted the drive as normal. Potentially quite exciting actually, because there were several low sections of road where water often flowed across the road at what we called 'inverts'. The water could be quite deep, lapping the bottom of the car. The water was on its way to one of the local swamps which for much of the fifties and sixties, and again in the mid-seventies were full of water for much of the year, if not the whole year.

The other thing about the drive on that Friday night was that we weren't on our own. There would be a string of taillights stretching out in front of us and looking behind, a similar string of headlights. Dozens of cars making the trip home from the traditional Friday in town. Consequently, if you didn't handle the road conditions well and failed to correct a sideways slide on a particularly wet and slippery section and ended up bogged in the 'table drain' beside the road, there would soon be a line of following cars pulled up to help. It would be all hands to the back of the bogged car to push, while various instructions were suggested for the driver to follow – like 'Turn your wheel left!", "Give 'er all you've got!" – and if you were right behind the wheels, you would be showered in mud as the wheels spun. If that failed, it would be time for someone to get out a tow rope, and back another car up. If that too failed, which I don't recall being very often, it was time to despatch someone to the nearest farm for a tractor and chain.

This level of inconvenience caused by poor roads was of course an issue of concern and was being rapidly addressed in the post war years. It was the era of building bitumen roads out from the major towns into the surrounding districts, and Nhill and the then Lowan Shire were part of this program.

Pretty soon, that one mile of bitumen that stopped at the golf course became three miles and was added to each year by at least another mile, and a whole four miles in one particular year to fix the potentially wettest section. As a result, in a fairly short period of time the drive home from Nhill became much less adventurous. And the road building program extended to more than the 'sealed' (bitumen) roads, but it also involved many more miles of dirt roads being converted to gravel. By the mid-sixties, a vastly improved network of roads had been built to make travel much more convenient throughout the district – as it happens - just as the series of wet years gave away to drier times.

There was another notable feature about the extension of the bitumen road network worthy of discussion here, and that was where some of them led to. Maybe it was just plain serendipity that a bitumen road seemed to go past, or lead to, the front gate of all of the Lowan Shire councillors who represented the three rural ridings within the shire!

To be fair, the first priority was obviously given to connecting the key localities such as Netherby, Lorquon, Yanac, Broughton, Diapur, Winiam and Woorak with Nhill. But once that was achieved, it did seem that providing a councillor with a good road was seen as a reasonable planning priority.

Now I have to declare that we were beneficiaries of this, as our Dad, CT Dodds, was a Lowan Shire councillor for twenty-seven years! As it happened, it was right towards the end of that twenty-seven years that the sealed road finally made its way to the corner of Perenna and Perenna School roads - our place. Maybe it was because of his propensity to always seek to please his constituents first, or most likely because ours was the furthest location from Nhill. Whatever, its arrival finally removed all the previous test of driving skill and challenge that the road to Nhill once was.

Set against today's standards of probity and governance expected of people in public office or other leadership positions, the building of a bitumen road to a councillor's door would seem questionable, but at the time it didn't resonate as an issue. In fact, it was openly discussed, often with light-hearted sarcasm, but there was general agreement that as a matter of principle, if you were a councillor then you couldn't properly discharge your duties unless you were able to leave your property and drive to Nhill for council meetings at any time of the year.

Probably not unreasonable! Therefore, if you gaze at a road map of the district now, the apparent illogical 'dead end' routes of some sealed roads might make more sense.

Dad did spend a long time as a councillor, and roads were always the major focus of the task. He seemingly spent countless hours on the phone at night talking to locals about the various roads that needed council's attention, as well as leaning over the bonnet of a neighbour's car in our driveway while discussing why a particular road should be graded or culvert fixed.

Roads were always a topical issue, even on a Sunday. One of the more memorable trips home from church was the day that Arnold Duthie, a local farmer, fellow Lorquon Cricket club member, and a Methodist 'lay preacher', led the Sunday Service. I recall Arnold's sermon on the day as one that made sense to me! In fact, I might still even be a regular church goer if every time I went, I wasn't preached to about God's wrath to be wrought upon us sinners but heard a well thought out message such as Arnold's that day. It was at a time of recent good rain, so he used the example of how he had to struggle to drive his car as he 'slithered and slid' on the slippery, muddy road up the hill towards his house, and used it as an example of how we often had to meet with and overcome difficulty.

I thought it was a nice use of metaphor, but Dad's response was different to mine, as we drove off from church. 'Fancy using the pulpit to complain about his road like that!'. "If that's all I'm going to hear, I won't be going back again". "We only had his road graded the other week, and that's all the thanks we get!"

Maybe it was coincidence, but it was probably only a year or so later that a nice new bitumen road went up straight past Arnold Duthie's front gate!

The Nhill Show

Show Day was one of the calendar events of the year in every town in Australia, no more so that the Annual Nhill Show. Pretty much the whole population turned out for the day – townsfolk as well as the farming community. Jack Dodds was an Agricultural & Pastoral (A&P) Society Committee member, under whose auspice the show was conducted. Accordingly, on show day he would set off early, dressed appropriately for the day and his importance – suit with waistcoat and tie, new hat, and very well shone shoes. Nanna of course would be also in her best outfit, with hat and gloves. We would follow later, also dressed in our Sunday best.

The show had all of the attractions – a huge exhibition of farm animals of every description, displays by local traders, side-show alley, a well patronised publican's booth, and of course the Ladies Exhibition – cakes and preserves galore.

There were many events for horses on the main showground, including trotting races, a highlight being the JW Dodds Handicap! Being not particularly into horses (maybe as an overreaction to the cricket pitch story earlier discussed) I might have stood by the fence to watch the trotters' trundle by in this particular event but left the rest of the horse events – of which there were many – to others.

As a young lad, my interest would be to have a quick look at the local traders' exhibits, which was quite large in these pre–Agricultural Machinery Field Days times. There would be a line-up of new cars of various makes to be admired, plus lots of new tractors and farm machines. At the time, there were several car dealerships in Nhill – at least four, all selling different makes. And in addition, there were at least the same number of machinery dealerships, all holding stock of the newest tractors available, plus harvesters and an array of other new equipment. All together these car and machinery dealerships were a significant part of the town's economy – all employing numbers of mechanics, sales and office staff. I recall that one such thriving business – Bullen Bros., the Holden Dealers was a bustling place, – with at least twenty staff. Just around the corner, Fishers Motors were the Ford Dealers with a similar scale of business, matched by Bongiorno Bros. with their showrooms and workshop, selling and servicing Dodges and Chryslers. And then there was the British Motors garage, where Austins and Morris cars could be acquired.

No such businesses are left in the town today, such has been the scale of change.

But back to the Show. After my quick circuit of the cars and machinery, it was off to Side-Show Alley. This was where all the young people hung out, and allowed ourselves to be relieved of whatever cash we may have had in our pockets. And what was the highlight? For the boys and men, it was the legendary Jimmy Sharman's Boxing Tent!

This was a very impressive big marquee, with a big platform above head height out the front, all draped with tarpaulins and banners. This is where the troupe of Boxers would stand, all fitted up with their boxing gloves, ready for action. From big intimidating looking heavyweights down to lightweights. It was a mix of circus, performance and the sport of boxing, which was very mainstream and popular at the time. Up on one end of the platform, one member of the troupe would stand beating a big brass drum, the percussion of which went right across the showground, probably scaring the heck out of all of the chooks on display over in the poultry pavilion in the far corner by the road. But then they were most likely used to it, as they would have been dragged around to several similar local shows in the quest for a First Prize Blue Ribbon "Best Laying Hen'. At the other end, stood the showman himself, Jimmy Sharman, with his megaphone who would be spruiking the reasons why you should line up to see the show. He would introduce each of his fighters and issue the challenge to locals "Who'll have a round or two for a pound or two?"

It is a matter of history that at least one of the fighters in the troupe would be aboriginal, and probably more. There were – or are – a couple of reasons for this. Firstly, as with many sports, they were very good boxers, or rather, sportsman in general. This, of course, was well demonstrated by several carving out international careers – such as Lionel Rose becoming a much-celebrated World Champion in the late 1960s. Secondly, it was one of the few ways indigenous folks at the time could go out and earn some real money while demonstrating their skill. At the time there is no doubt that racism and negative stereotypes about people of aboriginal descent were real and widespread, with opportunities for them very limited, therefore it was a plus that the Jimmy Sharman troupe provided an opportunity for some young aboriginal men to get ahead.[6]

Hence, the build-up for the show would continue, as Sharman explained the troupe's abilities and achievements, and sought out locals who thought they could take them on! This leads me to the day I most remember when the High School Headmaster's son put up his hand and said he would take on the young aboriginal lightweight! You can imagine the stir and interest this caused amongst his schoolmates, me included! We all paid up and filed into the tent to witness this contest. With us all whispering to each other: "What would Mr or Mrs Donovan be thinking about this?" "Do they know?" "He'll get knocked out for sure!" We looked around and couldn't sight our Headmaster anywhere, so assumed this challenge went ahead without his knowledge or blessing! The memory is faded of the outcome – but I do recall that our schoolmate was no match for his adversary, who took it pretty easy on him. The referee, Jimmy Sharman himself, let it run for a couple of rounds before he called it off. It was a well-earned 'Pound or Two' for our unexpected local combatant.

As the Show drew to a close, and the beat of Jimmy Sharman's drum fell silent after their last event of the day, many of the men gravitated towards the bar, which

---

6 Naranderra Argus, May 15, 2018

was a rotunda type of structure, with drinkers gathered all around. Here was the location of another showman's source of entertainment – a vertical pole about five metres high, with a bell on the top. The pole had a metal track running up one side, with the numbers 10-100 bottom to top. A little wheeled 'dolly' slotted into it so it could run up the track towards the bell, provided it was propelled by sufficient force, via a hinged catapult type wooden lever at the bottom which had a big pad of rubber tyre on it which needed to be struck by a sledgehammer, to launch the dolly with its marker skywards. Each hit with the sledgehammer was for a suitable fee of course, which as the beer flowed seemed to be an expense of lessening concern. Fuelled by the residual testosterone from the last half hour in the boxing tent, plus several beers, the young local blokes competed for the glory of ringing the bell. It was very entertaining, and revealed as with most things, there was an element of technique as well as sheer size and brute force, to the embarrassment of many and the applause of others as a slight little bloke kept on ringing the bell!

But there was more to the A & P Society than the conduct of its annual show back then – and I believe that this is true of the A & P movement across the country. Nowadays, they pretty much exist to conduct the local annual show, and little else. My understanding and recollection is that when they were originally formed they had a much wider scope of activity, which was to promote good agricultural practices and knowledge generally, and to encourage innovation. The Annual Show was a means by which the result of all this could be showcased. They represented all that was good and new in mainstream farming – agricultural and pastoral.

Beyond the Show, the A&P Society played an important role in the vital area of scientific research, and the extension of new cultivars and methods to the wider farming community.

Farm Field Days

One of my early memories of Fafa was hopping in the Ute with him, and heading off to a Field Day, on the Dahlenburg property at Salisbury, just east of Nhill. The A&P society's role in the research effort was in identifying property's that would be suitable for the trialling of new crop and pasture cultivars, and associated fertilizer and other experiments.

The conduct of the trial plots, and the plant breeding effort was all managed by the then Government Department of Agriculture. There was little or no involvement in agricultural research by private companies or organisation's back then – agriculture, like most industries was heavily regulated. Government agencies were charged with the task of research and development and funded accordingly. Some of the funding came from a research levy that was applied to commodities produced by farmers. This was all quite easily managed and applied at the time, because wheat in particular was totally regulated, with only one buyer of wheat – the Australian Wheat Board which was essentially a government agency. All wheat produced for sale went to the AWB, who managed payments to farmers for their product, and of course any levies or charges associated with the storage and marketing of the product.

The history of the regulation, and then deregulation, of most farming activity in Australia is a very interesting one, and in fact is a whole other story that I don't think belongs in this memoir, other than to provide background to my experience. It is relevant to my story of life on the farm, as the reach of regulation or government involvement in farming was at its peak through the 1940s, 1950s, and 1960s.

It is interesting to muse now about how this fitted into the politics of the time. As we moved into the Post World War II era, a strong theme was the threat of socialism, particularly as represented by the communist movement which had overtaken Russia, China, and much of Eastern Europe. All the evidence we saw from those countries was of government control over just about everything, with little individual freedom, particularly economic. The Government owned and controlled everything.

History now tells us that this resulted in a lower standard of living and loss of freedoms and rights of pretty much all citizenry apart from the ruling elite. All the evidence is that, within certain limits, free markets and free individual choice is fundamental to a successful society.

Yet, at the same time as we were all in total agreement about the threat of socialism taken to its extreme, our farming organisations were at the forefront of the quest to set up monopolistic government controlled or backed organisations such as the AWB, who were in total control of the marketing of commodities such as wheat. And there were no stronger advocates for this than Jack and Cliff Dodds, who would often relate how proud they were to have been part of the fight for 'wheat stabilisation', a government guaranteed price for wheat, and 'single desk' marketing. What this in effect meant is that through the AWB, the federal government 'owned' all of the wheat produced in Australia, and as it was sold, made payments to farmers out of one big 'pool', less costs of storage and marketing.

A closer look at the reasons behind this provides an insight into this apparent contradiction whereby fiercely independent farmers sought to and succeeded in establishing a communal system of marketing.

It seems to me that the impetus came from the highly volatile price of wheat, dating back to WW1, but particularly through the depression years of the 1930s, and again under the national interest of securing food supply during WW2. I don't doubt that Dad's often made claims that the grain traders used to 'rip off' the farmers with very low prices offered to purchase wheat during the depression years had at least a grain of truth about them - pardon the pun. Probably the best source of information that a farmer had with respect to market prices would have been something like the 'Weekly Times' newspaper, which as its name suggests was published weekly. A far cry from today, when a farmer can swipe his or her iPad or laptop and gain real time information on the value of every imaginable agricultural commodity. The traders held information that was not available to farmers at the time.

Their response was to agitate for a series of 'grower controlled' boards and agencies who had the capacity to compete in world grain markets on a more level footing with large often multinational companies. There was bipartisan agreement about the need for such intervention at a political level, so organisations such as the AWB developed and grew in influence as various pieces of legislation required to regulate and support the industry were enacted. Support was widespread across the industry as growers enjoyed more stable prices, generally allowing for a profit margin above the cost of production, and the likes of the AWB became almost 'sacred cows' amongst the farming community, lauded as they were as examples of what happens when 'grower control' is enabled.

I believe the success of this form of marketing during the boom post war years was probably due more to demand exceeding supply as economies recovered and grew, rather than any intrinsic advantage it might have had. The downside was that as a monopoly trader, the AWB was not subject to competition within its supply chain, with the potential for its cost of operation to rise with little challenge. As time went by, farmers became restive about the costs of freight, handling and storage which eroded the profit margin, particularly as world supply increased. A further issue that I came to find concerning about the model was that while having a grower-controlled board seemed to be a great idea, skill levels across the various disciplines of governance, financial control and technical and marketing experience were vital for those overseeing a multi-billion-dollar enterprise. Good blokes and good farmers as many board members undoubtedly were these skillsets required external appointments. In the end, I believe it was poor governance that brought about the demise of the AWB, particularly the so-called 'Iraqi Affair'.

My further musing about the dominant role of the AWB was that a generation of farmers (me included) were really divorced from the marketing of our product. We were merely suppliers of a bulk commodity, and as long as it met a basic 'Fair Average Quality' (FAQ), that was all we needed to do. Our goal became to produce as much FAQ wheat as we could, without necessarily being concerned about what the processors and end users thought of our product. Did my truckload of wheat make better flour or bread than any other? It didn't really matter. As the years went by, I came to realise that most really successful businesses achieved industry leadership by providing their customers with a better product (real or perceived) than their competitors, but as an individual grain grower, 'orderly marketing' meant that striving to have your product differentiated and seen to be more valuable than others was not possible.

The case could be argued, that at a national level, the AWB was able to offer a better product than others, as the Australian climate made our wheat 'the best in the world' So, did the collective gain balance off the individual loss? Maybe it did.

I do believe there was a benefit from this structure that took marketing away as a bedrock aspect of our business, and that was we became more specialised, focussed on how to produce high yielding crops. And importantly, this focus on crop production was science-based, informed by a system of agricultural research and extension, laudably strongly supported by government of the day. This leads me back to the Field Day at Dahlenburg's farm when I went along with Fafa.

This would have been in my formative high school days and fitted well with my interest in science – which I have maintained until this day.

The trial site was laid out almost exactly as such research sites are today, sixty years on. It occupied a rectangular area with the farmer's paddock, which was then divided into many 'replicated' plots, or rows of different species of wheat which were part of the Dept. of Agriculture's plant breeding work. This would include the varieties currently being used by district farmers – such as 'Pinnacle' or 'Quadrat' or 'Insignia', and it would also include one or more 'old' varieties no longer grown, such as 'Federation'. Most importantly, the trial would include several new varieties that didn't have a name as yet – just a series of letters and numbers as identifiers.

We would be met on site by Dept of Ag. Plant breeders and scientists, there in dust coats holding onto their clipboards (or was it just a notebook?) of information about the trial varieties, and the treatment applied to each plot in the trial. The main days we attended would be just prior to harvest with the plants fully out in ear, looking at their best. We would be taken around the site with the Dept of Ag people explaining the attributes of each species and its potential or drawbacks.

The main talking point would be which would have the best yield, and how much better the new varieties might look in comparison with the present and historic species, informed to some degree by the results of earlier glasshouse experiments that led to the field trial stage.

Of course, the results of the field trials would not be known until after harvest, when the Dept of Ag people would bring a little harvesting machine on to site (as plant breeders still do today), and carefully harvest, weigh, package and label the grain from each plot, before taking it back to their laboratories for further analysis, which included dough quality and baking tests.

Every year or two, a species which had shown yield and quality advantages over those existing would be identified and given a name, and one or more seasons then committed to 'bulking up' the seed so that it could be released to district farmers.

This whole process was prior to the present day where such work is undertaken by commercial organisations, who understandably require making a profit from the results of what is often many years from the original glasshouse cross breeding, through field trials, to eventual registration. Today a form of patent or Plant Variety Right (PVR) is held by the company which develops a new cultivar, and seed royalties flow to them. Seed for new species can be quite expensive, reflecting the effort and risk involved.

But back then, when we came home from the Field Day at Dahlenburgs, if we thought one of the soon to be named varieties would be a good one to grow, your best action would be to 'put your name down' with the Department in what was some kind of ballot, where, if your name was chosen, you would be offered three bags of the new variety as soon as sufficient seed was available.

Consequently, I can remember Dad very happily getting off the phone one day to announce that "we have just been allotted three bags of the new 'Olympic' wheat." We had been successful in the ballot! But this wasn't a licence to print money by becoming the holders of seed of this new species of wheat – particularly one named after the upcoming Olympic Games in Melbourne! It just gave you the chance to bulk up a supply of seed that would enable you to grow a decent acreage within a year or so, and the honour and pride in having this new crop on your farm!

With the three bags of the new seed wheat secured, the next step was to mark out a ten acre 'Land' on a fallow paddock earmarked for wheat. The ten acres was calculated based on the accepted seed rate when sowing a crop was a bushel of seed per acre. As there were three bushels to a bag, this equated to enough for nine acres, but given the desire to maximise seed production, the seed rate was adjusted downwards to spread the new seed over an additional acre – hence the ten acres.

The ten acres of new wheat on its marked out 'land', in a paddock adjacent to the main road, so that it would readily be able to be admired not just by us, but all the neighbours as well, was closely watched as the year progressed.

Come harvest time, the header would be carefully cleaned out so that the seed could be as pure as possible, and it was bagged out into new jute bags, which were sewn up and stacked away in the barn. If the yield came in around ten bags to the acre, which was generally considered to be a good result, you would have one hundred bags stored away. If the season was good, and the new variety yielded better than average then you might have one hundred and twenty or even one hundred and fifty bags. This would be more than you would be likely to use for the next crop, as it was good practice not to only plant one variety of wheat. One hundred bags of seed would be enough for three hundred acres, which would represent a quarter of the entire farm, so there would be excess seed for our needs. This would mean that we could provide neighbouring farms with a quantity of the new seed.

Given the regulated market of the time, all wheat was technically to be delivered to the AWB, unless you went through a process of gaining a permit to sell wheat privately. To avoid this bureaucratic process, and in keeping with the egalitarian nature of the time and the industry, where everyone received the same price for wheat, a process of 'swapping' any excess seed wheat you had for a similar quantity of' FAQ' wheat was followed. This obviously preceded the practice nowadays where a new variety of wheat is valued at multiples more than the market price for bulk wheat. I'm not sure, but maybe there was also an implicit or even explicit expectation that given you had secured this wheat via a ballot, then you were obligated to pass on excess production to others in order to speed its adoption across the region.

The way this was done was that if you agreed to provide a number of bushels of seed wheat to another grower – let's say thirty bushels, then he would come over and pick up ten bags from your supply, and would then ask the Officer in Charge at the local silo – where you all had to deliver your wheat – to transfer ownership of thirty bushels of wheat he had delivered into your name. No cash changed hands, and no profit was made from the transaction. It wouldn't happen today!

The Crop and Fallow Competition

The other role fulfilled back then by the A&P Society was its management of the annual 'Crop and Fallow' competition in the district. The theory behind this was that by entering into this annual event, local farmers would be able to demonstrate 'best practice' in the quest for good yielding crops, for the information and benefit of all. There was no cash prize, only a certificate which you could proudly hang on the wall or sit on the mantle-piece, if your crop was adjudged to be the best in its class. I cannot recall how many classes there may have been in which to enter, but we were occasionally successful in the 'Mallee Land' class. There might have been a 'Heavy Land' class, but we only had a relatively small area that fell into this definition, so our prowess was in Mallee Land!

As I recall, I think the minimum sized area you could nominate for your entry was forty acres (eighteen hectares).

To win the gong for the Best Wheat Crop (Mallee Land) you had to prepare for several visits from the august gentlemen of the Judging Panel. This sharpened you up, to make sure the paddock was looking its best every time a big black car rolled in the front gate with the judges on board. The judging panel, I recall, being two or three well regarded farmer members of the A & P Society, all suitably attired with suit coats and ties, plus a felt hat, along with an expert panel member, being a senior Dept. of Agriculture employee, with shirt and tie, and hound's tooth coat with leather elbows. If it hasn't already become apparent from reading this memoir so far, farming in particular, and society in general back in the fifties and sixties was a man's world. There was never a woman in that big black car! With respect to the car, it always seemed to me that if you enjoyed some success or status a nice big black car like a Dodge or a Humber, not an ordinary Holden or Falcon, would be a good way to demonstrate it.

The well-established and accepted method to reliably grow wheat in the district was an eighteen-month program of 'fallowing' the land for up to twelve months before you actually planted it, after which it would be a further six months before harvest. As the competition's name implied your success or otherwise was based on how good the judges thought your fallow was, prior to then having the subsequent crop subjected to their learned gaze.

As annual fallowing, as we practised it then, has become an almost completely abandoned wheat growing practice nowadays, perhaps I had better explain it. Essentially, the aim of a 'twelve-month fallow' was to save all of the previous year's rainfall as stored soil moisture, which your crop could then successfully utilise. And it worked. Crops sown on fallow land, with the technology and practices then available, almost always out-yielded crops sown on 'ley ground' or 'stubble' (meaning a crop or pasture had been on the land the year previous). This was certainly the case in years of average or below average rainfall. The primary benefit of fallowing was this stored moisture, although the reduction of soil borne disease was also a factor.

Fallowing was generally initiated in July. If the season was going okay, the paddocks to be fallowed would by then have quite a cover of pasture grasses and clover, which you would graze down heavily with sheep. A combination of compaction by the livestock, plus the binding together by plant roots meant that the first cultivation to commence the fallow would be the most difficult. Until the advent of larger more powerful tractors, this first operation was done with a mouldboard plough, an implement that would slice along below the soil surface, and invert the soil, burying the vegetation, and leaving bare earth on top. This demanded quite a bit of power, so with the tractors of the day – and previously horse teams, the implement was necessarily narrow. A six-furrow plough was a fairly common size of plough, which ploughed six feet (1.8 metres) per pass. Because you were completely inverting the soil, if you were lucky, your tractor might manage to do this at three miles per hour (5kmh). As a result, ploughing up two or three hundred acres of fallow was quite a time-consuming task. It would take more than half an hour to do the first 'round' around a one-hundred-acre paddock (40ha). I can remember taking a whole week to plough our 'Front' paddock the first year I worked on the farm. You could barely see what I had done after the first day! Thus, the farming calendar generally allowed a month to do the first pass of preparing your fallow – July into August.

Fallowing was a bit like painting the Sydney Harbour Bridge. It was a never-ending job, particularly if you wanted the paddock to be looking good in time for the first judging in the late spring. That being so, pretty much as soon as you had finished ploughing up the last of your fallow paddocks, there would be new weed growth on the first and it was time to begin the first of many passes of 'working the fallow'.

For this job, you backed the tractor up and hooked on the 'Sunduke' Scarifier and headed out to the paddock. The Scarifier was a different machine to the Plough, and about twice as wide at twelve feet (3.6metres). It had twenty-four cultivating tynes, spread between four or five rows fore to aft.

This first scarifying following the plough was probably one of the reasons so many of us farmers ended up with bad backs. The plough left lots of big lumpy clods of upturned soil, interspersed by the furrow lines – the soil surface was a sea of big bumps and clods. How well I remember dragging that scarifier across those really rough ploughed paddocks. The tractor bounced and bumped its way along, without the benefit of power steering to take the feedback back out of the steering wheel which kept trying to turn itself sideways, testing the muscles in your arms to the limit. And with no such thing as a proper suspension seat, or even padding on the metal seat other than an empty folded up wheat bag, (and of course no actual sprung suspension on the machine itself), you just banged and crashed your way along, trying to keep your backside properly located on the seat.

Thankfully, the effort was worth it, inasmuch as the scarifier did what its name suggests, and scarified this lumpy mess into a smoother more even surface of neater furrows. How you looked forward to getting to the middle of the paddock, and then being able to drive back across the newly levelled ground as you 'did the corners'

With each few weeks during the early spring, and with more rain and fresh weed growth, it was back out to work the fallow again with the scarifier. After a couple more passes the paddock would be broken down sufficiently to allow the use of a lighter and wider implement, a cultivator bar. If you could get it right, the last pass of the cultivator on the nominated paddock would be just a couple of days before the arrival of the judges. Hence, you would be focussed on keeping the tractor as straight as possible, leaving neat straight furrows that looked great as the sun sunk lower in the sky, leaving the bottom of the furrows in shade. And if you finished off with the 'headlands or corners nice and straight at a forty-five-degree angle, it looked a picture, as you drove out the gate and back home. It was one of the joys of farming to look across your beautifully cultivated paddock, and you hoped the judges would be similarly impressed by your workmanship.

On the day, they would climb out of the car and walk out into the paddock, comment suitably on how good the paddock looked, and scratch around to see if you had worked down to an even 'seed bed' of sufficient depth. As you can probably gather, much of the judging for the crop and fallow competition was a subjective thing – few if any measurements of anything were made, particularly as the crop was subsequently sown and grown through to harvest. The actual eventual yield of the crop was never asked for or expected. The judges used the same technique as a crop insurance assessor to estimate potential yield. - Therefore, the economics or the profitability of the wheat crop were not a direct subject of inquiry with respect to the best Wheat Crop Competition – it was largely a consideration of your workmanship.

I must admit this went across my developing belief that it was essential that farm profitability was elevated to greater importance than farming as a way of life, and so winning the A & P Society competition was of somewhat less importance than your end of year bottom line. Decades on I still have no doubt that successful farming is clearly linked to business acumen and financial management as a base requisite above making nice looking paddocks of fallow. But it also true that much has been lost as 'farming as a way of life' has receded, and taken many aspects of community benefit with it, particularly social contact and engagement.

But the Judges did bring valuable experience and knowledge with them – and did at the time I recall back in 1964 offer some sage advice:

"Well boys, it's a lovely paddock of fallow, although it might be a bit too fine for this time of year." "You need to have it a bit cloddy to stop the wind, just in case it stays dry next year. You don't want to have it drifting if it's a drought."

Figure 12 Winners!

Figure 11 A proudly displayed Certificate

They, of course, had experienced what we hadn't – the drought years of the 1930s and 1940s, with memories of the Mallee as a 'Dust Bowl'.

The lessons of these earlier drought years were destined to be re-learnt by us younger farmers, but not yet! Our experience had been of a series of unusually good wet years of the fifties and sixties. The records show that from 1945 through to 1966, rainfall was at or above long-term average almost without fail. The drought year of 1967 was the first reminder that maybe the earlier disproven theory of South Australian farmers of the 1880s that 'Rain follows the Plough' or that our modern mechanised farming somehow insured us against the scourge of drought was false. What proved it to be false beyond doubt was the extreme drought year of 1982, when we all first learnt about 'El Nino' as an influence on our weather.

Severe drought, as we were to learn, could lead to total crop failure, which was not only a financial disaster, but left the soil surface devoid of vegetation. This was made worse if the soil had been over cultivated and left too fine – highly prone to wind erosion.

But for now, I must recount with pride that Ken and I did win the 1965 A&P Society prize for' Best Wheat Crop Mallee Land' with our crop of 'Stockade' wheat that we subsequently grew on that paddock of fallow! And received a nice, framed certificate to hang on the dining room wall alongside several others previously won by Jack and Cliff!

Crop and Fallow competitions have now receded into history, like so many aspects of farming over recent decades, as discussed earlier via the view from Perenna Hill.

The desire to have a really good fallow phase was a really understandable farm practice for the time. It was a key part of the 'Four Year Rotation' that Jack and Cliff, and most of the district's farmers had settled on in the immediate post war period. I think it's worth a bit of detail about the farming system of the time, what was good about it, why and how it changed.

I must restate my conviction that the application of the results of science-based research and development is fundamental to the success of society at all levels. The rejection of evidence-based decision making, and the rise of the 'post truth' movement that has reached as high an office as the President of the United States of America, unfortunately with its adherents here, is I hope a short-term aberration in how modern society should manage and organise itself for the common good.

Maybe it goes back to the good fortune of having some good schoolteachers, particularly in the area of science and British History and the amount of reading material we always had at home, but my respect for the value of science runs deep.

And one of my early 'light bulb' lessons on science was the day we learnt about the Agrarian Revolution in Britain, and in particular of the findings of one Charles 'Turnip' Townshend.

I learnt that 'Turnip' Townshend questioned existing methods of strip cropping used by the peasant farming class working the land, where-by they grew narrow strips of wheat, oats and barley. He proposed that crop yields would benefit by the changing of this three-year rotation using similar plants (cereals) to a four-year rotation that involved a clover crop, on which animals could graze, followed by wheat, then oats or barley, then turnips.

Under this system, the clover could 'fix' nitrogen in the soil, which along with the livestock manure would fertilise the soil, which the cereal crops would benefit from, and then the turnips were also good for stock feed, but also being a different species, reduced the level of soil borne diseases that might affect wheat oats or barley.

This move to rotational farming along with livestock grazing, could only take place if the previously open estates of the landed gentry were 'enclosed', and formed into individual fields for each crop, with hedgerows, fences and walls separating them.

This adoption of new agricultural techniques was bad luck for the peasants who no longer had access to the big country estates to carry out their traditional strip farming. They found their way in increasing numbers into the towns and cities where new jobs were being created by technological change of the Industrial Revolution, depopulating rural areas in favour of growing towns and cities. While I am not comparing the many farming families who have left the Nhill district, including Mary and I and family, with the peasant class of England in the 1600s, it nonetheless underlines a very long-term world-wide trend – less and less people are actually living on and working the land to feed and clothe everyone else.

The result of this change to rotations led to greatly increased yields – in fact England became a significant exporter or trader of wheat due to the increased production.

Accordingly, the four-year rotation practised by Jack and Cliff, and indeed most of the district's farmers at the time was essentially sound. The key difference was driven by the different climate from the England that Turnip Townshend operated in – less rainfall, (not enough to grow turnips!) and longer hotter summers. Hence, the institution of the 'long fallow' to conserve a year's rainfall to smooth out variable yields.

This system of rotational cropping served us well in the post war years. It was actually as close as you could get to what today is regarded as 'Organic Farming.' The only 'non-natural' or chemical inputs were the use of superphosphate fertilizer, which really is a natural substance, phosphate rock or guano deposits, treated with sulphuric acid, and seed dressing or seed 'pickle'. There was no use of herbicides or pesticides – there was no such thing as a boom spray in the machinery shed.

Today's farming practices for broad acre cropping are very different – except for the few who practise Organic farming.

So what happened to bring about the replacement of one system of farming with another?

My answer is that, just as earlier farmers had followed the best science of the time (Eg. 'Turnip Townshend), today's farmers continue to do so, with an added understanding of the fragility of the Australian landscape and a challenging and changing climate.

There were some problems associated with the practices we followed in the fifties and sixties. Foremost was that bare fallowing left the land exposed to the threat of erosion, by both water and wind. Water erosion was bad for two reasons – firstly it washed your fertile topsoil away both by sheet and gully erosion, a bad enough outcome, but it also meant that the most limiting factor in the generally low rainfall landscape was made worse – less availability as subsoil moisture for the use of your crop. If at times of heavy rain most of the water ran away down a drain and ended up in a local swamp, then it was not there to benefit your crop towards the end of the year.

Figure 13 One of our very wet paddocks in 1988

Equally, wind erosion when severe could remove vast quantities of topsoil across whole paddocks. The 1982 drought was our 'wakeup call' about this threat. The day that a huge cloud of dust rolled over Melbourne in early 1983 was one I will never forget – sitting in the Ute a couple of miles north of home with visibility literally only a few metres, while thousands of acres of land that was sown to crops which had failed in the previous spring due to the drought, on both ours and neighbouring land, blew away in the heat and howling wind. I (we) had never witnessed anything like this before – it was a salutary lesson. Flying over the district and our own farm in a light plane a few weeks later further reinforced a sense in my mind that the accepted farming methods were no longer valid.

One of the key culprits that led to our soil being so prone to such erosion was a result of a key post war technological change in farming – mechanisation. We had moved from small and slow methods of cultivation to huge high horsepower tractors towing big wide cultivation machinery at speed, with repeated passes. These units could in total weigh up to twenty tonnes as they rolled across the paddock at ten kilometres per hour or more. The soil was pulverised by the implement's spring tynes and rolled flat by the machine's weight. This all led to soil structure decline and soil compaction. Exactly the opposite of what was needed for good plant growth.

There were other problems, again revealed by scientific understanding. There was an increasing awareness that our crop yields were lower than they should have been. This was based on the findings of a couple of South Australian researchers (French & Shultz) who devised a model of theoretical yield for dryland farming - expressed as taking aggregate growing season rainfall (April – October), subtracting estimated evaporation for that period (80 – 100 mm), leaving an available moisture profile to produce yield. The researchers then produced models of what each cultivar if properly managed should yield based on this available moisture. For wheat, this figure was twenty kilograms of grain per millimetre of moisture (20kg/mm). Based on long term rainfall records, this meant that across our whole farm at Perenna, we should have been achieving an average wheat yield of 3.2 tonnes per hectare. In reality, we were barely producing half of that, which meant lots of potential was not realised by our farming system.

As a consequence, we had to question why weren't our crops yielding to potential or at least much closer?

The answer is complex but could be simplified down to: failure to allow all of the received rainfall to infiltrate the soil and remain available for the crop, lack of all of the nutrients required, (principally nitrogen and phosphorus), and pests and diseases.

The good news story of farming in the Wimmera today is that many of these issues have been addressed, and crop yields are now much closer to the theoretical potential.

Backed by science, today's farmers have a much better understanding of what nutrients their crops require and how these can be made available, and a similar understanding of crop disease, for which plant breeding and agronomy have made big gains. But most important of all has been the move to zero tillage which in turn has enabled the retention of previous crops residue on the soil surface and in turn improved soil structure, and the infiltration of rainfall and its retention in the soil profile. The success of this system has also relied upon complete weed control from harvest time until the next season's seeding time, and earlier seeding allowing crops to have full use of the growing season. Perhaps controversially in the minds of some, the use of herbicides to achieve the necessary control of weeds in a zero-tillage system has also been a key to the much higher yields now being achieved.

As a result, today's farming methods are vastly different to those that won the approval and admiration of the Judges of the A & P Society's Crop and Fallow competition. And I am sure that the gentlemen of the judging panel who gave weight to straight driving and straight furrows across a paddock of fallow would be astonished with the GPS guided perfectly straight lines of newly emerging crops across literally tens of thousands of hectares of crops across the district today – where the tractor driver doesn't hold the steering wheel at all.

All that technology has further reduced the farm labour required, and therefore the continued population decline beyond the major regional towns. So, there is greater crop productivity, but far less people involved.

The Railway Yard

The Netherby railway yard, along with hundreds of others across the country, is today a deserted place; it is most unlikely that you would ever come across another person there at any time of day. The railway line itself is no more – the last train that ran to Netherby was in 1986, and soon after all the rails and timber sleepers were removed and used elsewhere, or for firewood. The station platforms, goods sheds along with a crane to assist with loading or unloading and the station master's office are all gone. All recycled or re-purposed to use today's vernacular. So too is the house in which the station master and his family lived, plus another three dwellings for the railway "gangers" and their families. Still standing are the rather substantial but now unused grain silos and grain shed. Maybe they are awaiting their turn to become canvasses for more 'Silo Art', a new and interesting initiative that is drawing visitors back to places like Netherby.

The is no better example of how rural communities have changed in just a few short decades as the now abandoned railway yards. They were such an important part of our lives. It wasn't just a place to drop off that cream can. It is hard to imagine now just how busy a place it was.

The most important role the railway played for us, and in fact the one that lasted the longest before its final demise, was as the point of delivery for all of the grain produced in the district. This made it the focus for us for the grain harvest period

Figure 14 Unloading Wheat at Netherby

– which during the 1950s and 1960s used to last from mid-November right through to the end of January. This became more compressed as bulk handling and higher capacity harvesting machinery were introduced – this increasing development of technology eventually rendering the facilities at Netherby obsolete, as the capacity to receive grain was simply insufficient.

The concrete silo, like many others built right across the grain growing areas of the 1930s would have been state of the art back then, with a capacity to easily handle the grain harvest of the times. The harvesting machines could work all day, and still not process the tonnage of wheat that today's machines can easily achieve in an hour. And five tonnes of wheat on one of the farm trucks of the day was a really good load – compared with fifty tonnes today. A new way to handle large volumes of grain in a short time had to be developed – and the answer to that challenge was not at Netherby, or indeed at hundreds of similar sites across the grain belt. They are now silent reminders of a different time.

How different it was. By half past six in the morning during harvest time back then, the first of up to fifty trucks would be changing down a gear as they turned past the pub on the corner, for the run downhill to join the queue parked waiting at the sample stand, located just before the weigh bridge. Silo hours were eight o'clock to five o'clock with an hour off for lunch at midday, so by arriving by seven o'clock to get your place in the line-up meant at least an hour to sit around in groups, having a good old social catch up.

Harvest really was a more relaxed time then than now. It was a given that it would be spread over a two month period from late November to late January, with an unofficial target to finish just before Australia Day, so that the cricket tragics, of whom there were many , myself included, could get over to Adelaide for the traditional Australia Day Test Match, with the family ensconced at West Beach or Henley Beach for a fortnight's holiday by the sea. So, at the time there was little questioning of the eight hours only operation time for the silo, including a lunch break. And of course, at the weekend it was only open on Saturday morning, because cricket and tennis were a higher priority, and there was never any consideration of opening on Sunday and breaking the Sabbath.

Of course, the main topic of the morning conversation was always how good everyone's crops were going, but soon spread to other issues. And like at church on Sunday mornings, anyone who might be missing was quickly noted, and the reasons for his absence proffered. 'Yeah, I think he broke the Knife Drive yesterday arvo, and had to go into town for some spare parts" or "He's in his back paddock, and it's only going a few bags, so he probably didn't get another load ready before dark".

A break in the general discussions was afforded by watching the railway 'gangers' loading up their railway 'trikes', before heading out along the line for a day of railway maintenance. The railway was a significant employer – at the time there were three 'gangers', who lived in railway houses just across the line from the silo with their families. There was a fourth slightly more impressive railway house, that of the station master and his family.

The Station Master, Len Decker, was one of several local identities who with their families made up the core of the busy and diverse little township that I recall of the early 1960s. Len seemed to be well fitted to his role, always suitably officious and taking care of details to a fault. The main interaction we had with him would be via a phone call to inform us that something of ours had arrived on the weekly goods train. This might have been as simple as an empty cream can, through to sixteen tonnes of 'superphosphate'. It might also have been a piece of equipment or other farm needs. Like the day the new steel cattle ramp for our front gate arrived. In this case we were summoned by Len to be at the railway yards at a certain time when the railway wagon it was in could be manoeuvred under the goods shed crane.

This was a scheduling task made to order for Len to show his organisational expertise! When the train arrived at Netherby, there were several wagons to be unhitched and left there, before the train choofed its way off towards Yanac, the last stop on the line. Of the wagons left at Netherby, several would be empty 'GY' wagons, to be loaded with bulk wheat from the silo. Thus, Len would have the Officer in Charge of the silo, Mick Stevens, who also worked as a local Stock and Station Agent, ready to go and have the wagons filled, each being manoeuvred under the outlet spout in turn.

The wagons were moved by manual labour, achieved by using a big heavy steel 'Pinch Bar' which you would use as a lever under the wagon's steel wheel, pressing down against the rail line to get the wagon to move ever so slightly at first, helped, if possible, by any spare onlookers who could assist by pushing. Once it was on the move, you needed to be ready to drop a big heavy brake lever mounted on the side of the wagon to bring it to a stop in just the right place. As Mick filled the wagon, he had to be ready to turn off the flow of grain, jump down and move the wagon forward a few feet at a time to enable the full sixteen tonnes of grain to be evenly loaded. Of course, as the tonnage built up, the manual effort required also increased. Mick had a busy day, because the train would be making its return journey back from Yanac later in the afternoon, and if he had his 'Rake' of ten GY wagons full and tarped over ready to go, they could be hitched up again, and off towards Jeparit and Dimboola on their way to Geelong or Portland to be loaded aboard a ship for export[7]. If for whatever reason he didn't get the job done, they would have to wait till the next week.

Meanwhile, Stationmaster Len would also have notified the local General Store owner Con Deckert which goods van his supplies were in, and most importantly,

7 A large percentage of the Wimmera grain harvest was exported through the 1950s and 1960s. While export is still important today, a significant portion of grain is now consumed domestically – largely for stock feed purposes. This has been due to Australia's population growth from around ten million then to twenty-five million, with the resultant increase in local demand. Large quantities of grain are now used by intensive livestock industries such as poultry, beef and dairy, as well as direct human consumption. The flow of grain beyond the farm gate is now much more complex, for which the system of silos all connected to big export terminals by rail was not necessarily designed.

would also have advised the Publican that his several barrels of beer needed to be unloaded as well. Anyone else who had any items of freight were advised by Len as to which carriage they were in to be picked up, such as our metal cattle grid.

Probably the main inwards freight that arrived by the weekly train was all of the fertilizer, or 'superphosphate' that would be used by the district's farmers as they sowed the new season's crop. Superphosphate is a rather heavy material, all loaded in 180-pound (75kg) bags and stacked into 'GY' rail wagons. As a result, it was a busy day or two for the Bedford truck when our 'super' arrived. It was all handled manually using 'bag trucks', or two-wheel hand trolleys, to load the truck, take it home, and then wheel it into the barn until it was needed at cropping time.

The process was to identify with Len's guidance which wagon or wagons held your super, and untie and roll back the heavy tarpaulin covers, and then use a hammer to bang open the big heavy door latches and swing them wide open, so that you could back the truck up to the side of the wagon, and get a rather heavy (it needed two men to carry it!) steel and wooden ramp from the goods shed to lay between the wagon and the back of the truck.

I have never quite established why we didn't upgrade our fleet of two 'bag trucks' which we used to load and unload all these bags. One was a 'Mackay' iron bag truck. It must have been designed in the early stages of the Industrial Revolution. It was made of thick steel tubing and quarter inch flat steel, a thick steel axle with steel wheels mounted on metal-to-metal 'flange' bearings. The thing must have weighed half as much as a bag of super itself, so even to push it along without a bag on it was quite a task. I expect this item must have been one of Jack Dodds's earliest purchases. The second bag truck was a snappy post war unit, made of much thinner tubular steel, with pneumatic wheels and real roller bearings. This was an absolute breeze to push along! It was an unwritten rule that we would swap trollies every ten bags or so. For some reason, a second similar easy-to-use bag truck never seemed to be allowed for in the annual farm budget! Maybe it was because of our belief that bulk handling was just around the corner, and the bag trucks would become obsolete, candidates to go 'down the well'.

As it happens, that wasn't their fate. Those bags trucks remained for decades more, and it is very fitting that Mary came across a replica of the old steel 'Mackay' and it now serves as an indestructible garden ornament at 'Mooringa' (Our home at Horsham after we left the farm) – far and away its best and most suitable application!

Taking out a Gym Membership in order to have sufficient 'weight bearing' exercise was at least forty years into the future for me, and similarly would have been a laughable concept amongst the gathered throng waiting for the weighbridge to open! Because wheeling bags of superphosphate was one of the less demanding tasks back then.

There was a perfect example of that just across from the weighbridge – the Barley Stack. Right up until the 1960s, the only grain received 'in bulk' at the railway yard was wheat - which went into the silo. Other grains, mostly barley, were still bought and sold, and handled in bags. These were stacked in the station yard, ready to be loaded into rail wagons in the months following harvest.

Consequently, apart from the weighbridge keeper and the OIC (Officer in Charge) of the silo, the other local harvest staff who turned up for opening time were the 'Barley Lumpers' These were three blokes whose task it was to start up the engine on a movable elevator, which was located so that a truckload of bagged barley could be backed up to it, with the top of the elevator positioned in the middle of the Barley Stack, which as harvest progressed would have grown to thousands of bags. Now, the good thing for the barley lumpers was that barley was a lighter commodity per volume than wheat, so a 'Three Bushell' bag of barley only weighed a mere 150 pounds! (65 KG).

The process was that the farmer (and any helpers he might have) tipped his bags off the back of his truck onto the start of the elevator, which conveyed it up to a point above the stack, about shoulder high above its current level. The 'Lumpers' would in turn line up and allow the bag to fall off the end of the machine to land across their shoulders, grabbing each end of the bag with both arms as they balanced, and then strode across the top of the stack to drop it into position. This followed a strict layer-by-layer plan, weaving it together like an interlocking stack of bricks. And that's what they did all day! A beer (or two) at the pub each night was well earned! The end result was a most impressive big rectangular stack, even more imposing than Jack Dodds' haystack mentioned earlier in this tale.

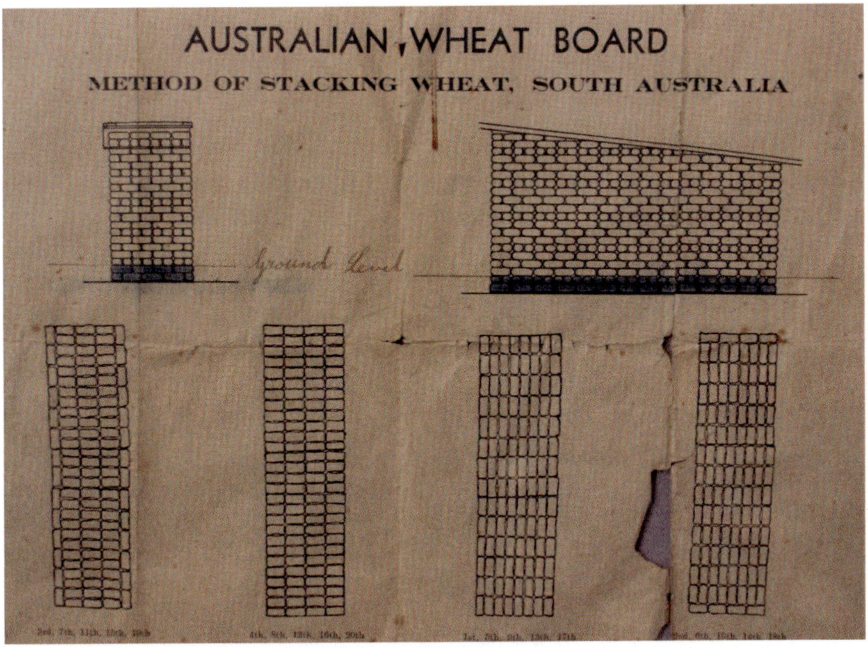

Figure 15 How to build a wheat (or barley) stack

That being so, to focus back on the harvest time silo line up, other distractions often provided entertainment.

The queuing protocol was as soon as the first few truckloads of wheat were weighed over to head up and tip off their load at the silo, but that any load of bagged barley, whatever its place in the queue would be waved through and weighed so that the Lumpers could get their day underway. Hence, the farmer's truck would be backed up, and he would proceed to drag the bags across the tray and drop them onto the elevator. Almost certainly a couple of characters would say "Let's go and give Bill a hand to unload and see if we can make these buggers earn their money!" So, they would jump up on Bill's truck and start feeding bags onto the elevator as fast as they could, yelling out "C'mon!" "Get stuck into it up there and shift these bags!" They would be laying them on the elevator practically touching each other end to end. The engine would start to labour as it struggled to move the weight, and soon the bags would be dropping off at the top, one after another. The blokes up on the stack rose to the challenge, wanting to make sure that their reputation as highly efficient Bag Lumpers stayed intact. They would be shouldering the bags and running across the stack to place it, racing back just in time to follow their workmates as they in turn caught the next incoming bag, yelling out "Come on you blokes down there, can't you feed them up to us a bit faster than that?"

Of course, this didn't last long, as everyone soon tired out, as well as the likely call - "righto you blokes, stop messing around over there and come and move your trucks up – the queue has to move along"

One of the vehicles in the queue was Ardie Nuske's Lanz Bulldog tractor, towing a trailer with its bin full of wheat. Lanz Bulldog's were a rather unique German built machine, which featured a large single cylinder slow revving engine, its design featuring a large and heavy steel flywheel mounted on its side. It was a very simple and apparently robust early example of an internal combustion engine and did not have such an advanced feature as a self-starter. Starting this machine required a bit of physical effort – you had to grab hold of the flywheel turn it back till just against the compression stroke, then quickly fling it around with enough force that on its next rotation against compression it would fire up and spring into motion, with a slow but very audible 'Thump' Thump Thump' exhaust note, with the flywheel maintaining the engine's momentum.

There was much amusement one morning when it was Ardie's turn to move forward towards the weigh bridge. One of the real characters of the district was Kevin or Joe Rethus. Joe was a tall ample framed, strapping big bloke, tipping the scales well north of 100kg. Joe was an obvious candidate to offer to help start Ardie's Bulldog up. It was a piece of cake for Joe – he grabbed the flywheel with Ardie staying in the driver's seat, turned it back, gave it one big heave and off the motor went - Choof Choof Choof!.

But there was consternation in the driver's seat! Ardie said to Joe, in his quite distinctive German accent: "Joe, you haff started it der vrong vay! It's going backvards!"

And it was – the engine was spinning the wrong way! Now there was nothing particularly bad about this, as the motor was happily idling away, awaiting its chance to be revved up and put to work. The only problem was that Ardie's machine now had one forward and five reverse gears! It was going to be a slow trip home after unloading! These simple old motors would really run either way – it just depended which way you turned it in starting it up.

The problem was quickly rectified, with Ardie turning the engine off and then asking Joe to swing the flywheel the other way. Another big effortless heave by Joe, this time in the right direction, and off went the Bulldog and its trailer to the weigh bridge.

The Powerhouse

It seems to be an historic fact that the early citizens of Nhill were an innovative, progressive lot. The record shows that the town developed quickly in the final decades of the 1800s. One of the interesting facts is that Nhill lays claim to being the first town in Victoria to complete the installation of a town wide electricity supply, way back in 1891[8]. Two-hundred-and-forty-volt electricity was generated at a Powerhouse, located just north of Nelson St, between where the Commercial Hotel now stands and the Bowling Greens, and it fed into a grid of poles and wires throughout the town, and in later years, out to the Aerodrome. But no further. The district's farms and other little townships carried on without electricity, before they in turn made their own what we would call today 'Off the Grid" arrangements.

However, the powerhouse I want to talk about here is in fact the small little gable-roofed tin shed that housed our thirty-two-volt power plant. It was a modest little building, with a wooden door at its front, located quite close to the house, to which it was connected by a couple of insulated wires. It was just outside the gate from where a path led to our back door, at the end of the driveway beside the house.

As mentioned earlier, one of the really beneficial things about its location was that it made a perfect backstop for our cricket pitch, filling in for a wicketkeeper. Therefore, by the time I was in my early teens, the tin on the front of the shed was looking a bit dented and less than pristine, as was the wooden door, which also made an excellent target for practising stab kicks and drop punts during the footy season.

Inside the little shed a 'Cooper' Petrol engine was mounted on a concrete block, with drive belts attached to a generator. Along the side wall, mounted on a timber frame were lots of big heavy batteries all connected together, ensuring there was power fed into the house if the engine stopped, or we hadn't bothered to start it yet. Underneath the batteries, a whole row of about twenty, one- gallon petrol tins were kept, for the purpose of refuelling the engine as required. And beside these, two or three earthenware containers that held 'distilled" water for topping up the batteries.

Refilling the petrol tins was one of the 'odd jobs' that I was given responsibility for. As soon as several were empty, it was my task to cart them over to the Machinery Shed where several Forty-Four-gallon drums of Petrol and Kerosene were kept, fill them up, and cart them back and place them back in the shed. I'm mentioning this task mainly to point out that there were all these little tins, some with their lids not necessarily screwed back on properly, with their associated fumes leaking out, just inside the shed door.

---

8 Research Data Australia

Filling the distilled water bottles was a more opportunistic activity. I used to ask exactly why we did it the way we did, and not just turn on the tap at the rainwater tank and fill them up there. Because what we used was not actually water specially distilled by some process – it was just rainwater. But Dad assured us, for the good of the batteries, you mustn't use tank water because it would be contaminated by the metal walls of the tank. So that was out. It never dawned on us, that this was all the water we had to drink, so why would it not be harmful to us also? Maybe it was!

Anyway, the way we got our distilled water (actually pure rainwater) for the batteries, was to wait till you saw a nice big black cloud looming in the west, and race into the kitchen and raid every cupboard to find all available pudding bowls, mixing bowls, salad bowls and any other earthenware, glass or china receptacle you could – anything but tin. Out through the back door and spread those all out across the small little patch of lawn there. And wait for the rain.

In those good wet years, it generally did rain when a black cloud appeared, so the task then was to decant all the water caught into a glass jug and fill the 'Distilled Water' containers in the powerhouse, leaving a big stack of bowls on the kitchen bench for Mum to put away.

One other thing to mention about the powerhouse was the rusty nail, hammered half-way into the wooden door, centred towards its top. Its purpose? It was there for Guy Fawkes Night! More about that later.

The thirty-two-volt power plant provided us with a reliable supply of electricity, which was mainly used for lighting. There were very few electrical appliances made for thirty-two-volt electricity – of which the main one was an electric iron. I think the washing machine must also have been thirty-two volt. Any other devices required a 240-volt transformer – in particular, our first TV, which provided quite a drag on that little Cooper engine and generator.

The most memorable inconvenience of this was what happened to the television picture if Mum happened to turn on the iron to press our clothes for school the next morning. There we would be, all sitting watching the grainy black and white image of the latest gripping episode of 'Rawhide' or 'Perry Mason' and the picture would shrink down to half size! And the light would grow dim. We would yell out "Mum, turn off the Iron!", to which she would reply "I have to get your clothes done, or you will look terrible at school tomorrow." Grudgingly noting this, we would crawl across the floor to get up close enough to see the now miniature TV image.

That all changed in 1966, when the State Electricity Commission supply, with its hundreds, or thousands, of miles of new power lines was first connected to Nhill's supply. This led to the decommissioning of the Power Station there. Our little powerhouse also became redundant, as we upgraded ourselves with all kinds of new two hundred- and forty-volt electric lights and devices. No more shrunken TV images!

Mallee Roots

One of the things that you certainly couldn't do with 32-volt power was run an Electric Bar heater, so that was one of the earliest new purchases for the house after the advent of 240-volt mains power, plus an Arc Welder for the farm workshop.

Prior to then, all the warmth that we had through the colder months of the year, in that highly uninsulated farmhouse, came from burning timber fuel. We always maintained two good big piles of wood, just out behind the powerhouse. One was a pile of 'Mallee Roots', or tree stumps, and the other was a big pile of foot long blocks, produced by sawing up various logs, tree branches and old fence posts that were collected from around the farm.

All of this fuel was fed into a Slow Combustion Stove in the kitchen, a big open fireplace in the lounge, and a hot water service or furnace out the back of the house.

Mallee Roots were a by-product of land clearing. Mallee trees, represented by several different genera of the Eucalypt species, were the very well adapted trees that dominated the Mallee Woodlands which was the original vegetation that covered most of our farm, and a large part of the surrounding district. And a very large part of inland southern Australia.

Mallee trees are multi - stemmed, with these stems all arising from a large 'lignotuber' (which was referred to as the 'Mallee Root') just below the soil surface. This particular adaptation allowed the trees to suffer significant damage, generally from fire, but to quickly spring back to life by producing new stems from this below ground source of energy, even though the top might have been completely destroyed.

These trees developed this way of surviving, having spent millions of years evolving through countless episodes of fire, until along we came with our axes, 'Mallee rollers', and finally bulldozers, with which we removed the tree stems. So, they were up for our first line of attack, by just growing up again. Therefore, we had to get more serious and kill off these lignotubers, variously by physically loosening them from their attached root systems, or by constantly rolling and/or burning them till they finally gave up and died and dried out. This left a paddock with hundreds or thousands of dead 'stumps' or mallee roots just below the surface. The next part of the land clearing process was to use various methods of ploughing to bring these all to the surface, allowing them to be physically picked up and carted away, or actually heaped and burnt on site.

They must have been, very early on, identified as a good fuel source. While they varied in size, they were mostly able to split into pieces with an axe. Their wood was quite dense, and with a gnarly surface could catch alight quite easily and burn hot.

Most, if not all, of the Mallee roots on our farm was long gone when I was young, but there were still thousands of acres of land being cleared locally, along the southern fringe of the Big Desert, and the 'Propadollah Scrub'. As a result, for many years, right up until the eventual enactment of land clearing laws brought a halt to the era of mass 'land development' in the 1980s, there was an ample supply of this source of fuel in the district.

Picking Stumps was another labour-intensive task that helped to keep us fit! They all had to be picked up individually and thrown up onto the tray of a truck, where they were stacked in such a way that when a full load was achieved, they wouldn't all fall off as you lurched out the gateway onto the road. It was task that was well beyond the capability of an individual farmer to do on his own. To make any sort of impact, you needed someone to drive the truck slowly along, at least one person up on the tray carefully stacking them all in an interwoven fashion as the load grew higher, and at least two or three on the ground throwing them up onto the load.

As a consequence, the few who were still clearing this land had to look to others to provide sufficient labour to make an impact on moving the stumps from the paddock. They would use a 'stump rake' to rake them into some sort of row every few metres across the paddock, and then seek the interest of a range of local sporting, charitable and service clubs whose members would volunteer their time to 'go picking stumps' and then sell them on a tonnage basis throughout the town and district, or even load them onto rail trucks to be despatched to a fuel merchant further afield. That's how we got to keep adding to our stump heap – by both volunteering to pick stumps and then to take a load ourselves.

The Party Line

One of my 'Deja Vu' moments was a decade or so ago when I successfully figured out how to bring another person into a phone call on my newly acquired iPhone. It was magic – you could have a three-way conversation, or even 'conference' calls with several people all at once! The wonders of modern technology! Then, a strange feeling that I had seen this before - I realised this was pretty much how our 'party line' phone worked back in the 1950s.

Connection to the outside world via telephone was a technology that was well established throughout the district when I was young. Telephone lines ran beside pretty much every road in the district. These were mostly just a single wire, strung between rough timber poles, generally built just inside the fence line of the adjoining farm. These lines were all connected back to the Lorquon Post Office and Telephone exchange. I could argue that they represented a good example of infrastructure built under the 'PPP' or Public/Private Partnership model that is in vogue for large infrastructure works today. We just hadn't coined the acronym.

How it worked, was that for the first few miles radiating out from the local Post Office/Phone Exchange, the government (via the then 'PMG' or Post Master General's Office) built and maintained the phone lines, after which the lines became the responsibility of the more distant landholders to build and maintain, if they wished to be connected.

In our case, the Lorquon exchange was ten miles away. For the first mile or so, our line, the '34 Line', (pronounced 'Three Four') was part of an underground cable. It then commenced its above ground journey, along with several others, attached to insulators on cross-arms, as a 'Main Line' This was the bit that was the PMG's responsibility. As this line passed by an adjoining farm, a single wire line, built by the landholder, would branch off to the farmhouse. As a result, the main line had less wires on its cross bars as it headed further out. At a certain point, this main line would come to an end, from where it was up to landholder's further out to build and maintain it themselves. This is where the 'Party Line' came in. A party line was a single line that was shared by several customers – something like five or six. Hence, as was the case out along the Perenna Rd., there were two single lines – the 'Three Four' line and the 'Double One' line, between them serving ten or twelve families.

The upside of this system was its simplicity and low cost – the downside, if it mattered, was lack of privacy, because everyone on the line could pick up their receiver and 'listen in' to any conversation on the line and join in if you felt like it. Just like a multi person iPhone call!

There was of course a protocol that was mostly followed. This was that you only picked up and answered the phone if it was your unique ring. Again, shades of today's smart phone – personalised ring tones! Our 1950s ring tones were based on a 'Morse Code' system which converted the letters of the alphabet into unique combinations of short or long pulses of sound, or 'dots' and 'dashes', which we referred to as 'shorts' or 'long'.

THE PHONETIC ALPHABET MORSE CODE GUIDE

A ALPHA	•–	N NOVEMBER	–•
B BRAVO	–•••	O OSCAR	–––
C CHARLIE	–•–•	P PAPA	•––•
D DELTA	–••	Q QUEBEC	––•–
E ECHO	•	R ROMEO	•–•
F FOXTROT	••–•	S SIERRA	•••
G GOLF	––•	T TANGO	–
H HOTEL	••••	U UNIFORM	••–
I INDIA	••	V VICTOR	•••–
J JULIET	•–––	W WHISKEY	•––
K KILO	–•	X X-RAY	–••–
L LIMA	•–••	Y YANKEE	–•––
M MIKE	––	Z ZULU	––••

Figure 16 Morse Code Chart

Our phone number was 'Lorquon 34S', which meant that if we heard three quick rings, or three 'shorts', one after another, it was a call for us. So, when the phone began to ring you had to concentrate. Was it 'Short Short Short' (34S- ours), or calls to neighbours such as Long Short Short, (34D), or Short Short Long, (34U), or Short Short Short Short (34H)?

The way you made a call was to turn a handle on the side of the phone – which in turn caused the bells on all of the phones on the party line to ring. If say you wanted to call 34D, you gave the handle one long and then two quick turns. If you wished to make a call beyond the party line, you turned the handle around for several times – producing one long ring. This call would be answered by the Postmaster, or his wife, at the Post Office in Lorquon. This would generally lead to a conversation as to "how was everyone today?", and "how much rain did you get last night?", etcetera, (there wasn't much going on in the district that the Postmaster and his wife didn't know about!) until they asked, "So, who do you want to be connected to?" You could give them either a name or number, and they would then plug your line into another local line if your call was local, or in turn call the main Nhill telephone exchange to be connected to the world beyond Lorquon.

This manual system involving having actual people making the connections 'down the line' for you lasted until the late 1960s when the wonder of an Automatic Telephone Exchange arrived. This change brought about the end of the party line system, and in fact most of the above ground phone lines in the district, which were replaced by underground cables. We all graduated to phones with rotary dials on them, allowing you to call direct, with no need for any local person to assist. With that, a key role of the local Postmaster, and the exchange staff at Nhill became redundant. The Lorquon Post Office itself closed soon after and in turn the adjacent General Store.

The other change that came from the introduction of this new phone system was that we no longer had to maintain the phone line ourselves. This had been a reasonably regular part of the farm calendar. Given that most of the roadside along which the phone line ran was also lined with remnant vegetation (most of vegetation left in the district was along roadsides), tree growth touching the line, or broken branches after a windstorm, often caused a 'short circuit' or phone outage. A big broken branch would be enough to pull the line down off its insulators, ending up on the ground, or lying across a fence line.

A dead phone line meant you got in the Ute, loaded up with a ladder, spare wire, wire strainers, pliers, a crosscut saw, (there was no such thing on the farm as a chain saw until the 1960s) and an axe, and headed off down the road, to see if you could find the trouble. The task was to get the line back onto its insulators, free of any tree limbs or other foreign objects. Then head back home and see if your work was successful by ringing up the postmaster. If he answered, it was fixed!

Jack and Cliff, along with other neighbours on the party line saw that it was a sensible precaution to do some pre-emptive maintenance along the line during the year, being to occasionally cut back some of the tree growth along its route. Most of the trees were Mallee species, with their multiple stems, so the tree trimming usually involved cutting back one or more of the stems that were closest to the line. Without the benefit of chain saws or the like, this was laborious, slow work, mostly by using an axe to chop through the stem near ground level. It didn't necessarily result in any wholesale destruction of many trees, just a constant chipping away at the task through the year, heaping the cut branches up in the roadside 'formation' or table-drain, to be burnt as they dried out. Until about the end of September, when an additional action was added. This involved taking the Bedford Truck down the road as well as the Ute, and instead of heaping the branches in the table drain, you loaded them on the truck and took them off back home. Why? Because it was nearly Guy Fawkes Night!

Guy Fawkes Night

What with various sporting events, (mainly footy and cricket) Church on Sunday, Fridays in Nhill, School Activities, we did have an active social life, to which were added some special events – one of which was Guy Fawkes Night. This fell on November 5 every year, apparently to commemorate a failed attempt by a bloke named Guy Fawkes to blow up the House of Lords in London with a heap of gunpowder. So, true to our British roots, we celebrated this thwarted plot each year on November 5 by blowing things up with gunpowder! Well, sort of.

We looked forward to Guy Fawkes Night each year for weeks in advance. The local tradition seemed to be that you rotated hosting a bonfire at your place with one or two neighbouring families. When it was our turn, the Bedford truck was busy for weeks before, carting in piles of tree branches and tipping them on to an ever-growing bonfire between the house and the machinery shed. And while this was going on, every trip to Nhill involved a visit to the fireworks section in Whiteheads Stores to add to a growing cache of assorted fireworks, which we safely stored in cardboard boxes under our beds.

Come the night, the neighbours would arrive with their big boxes of fireworks to add to ours. We would all have a meal together waiting for it to get dark – we didn't have to wait too long, because there was no daylight saving – and out we would go and light up the fire.

Then the fun began. With the light provided by the bonfire it must have taken an hour or so to blow up all the crackers or 'bungers', while watching a dazzling display of multi-coloured sparks that erupted like mini volcanos from 'Roman Candles', burning for a minute or more before they exhausted themselves. And the main attraction was letting off the 'Skyrockets'. You mounted these on, or in, the launch pad, an empty beer bottle, lit the wick, stood well back and waited.

This all predated the space race, so knowing nothing about countdowns, we didn't all stand around saying "Ten, Nine, Eight............". We just waited in silence, so that you could hear the hissing sound of the burning wick. And then with a whoosh it took off high into the dark night sky, before it exploded at the top of its flight in a shower of coloured sparks. Then of course, gravity kicked in, and the smouldering remnants of the spent rocket – a half burnt stick and blown apart paper casing, fell back to earth. These would land anywhere within a hundred metres or more, but we didn't particularly care. We were too busy lighting up the next one. Sometimes the noise of the landing wreckage indicated it had landed with a bang on the roof of the house, or the machinery shed. Otherwise, any issue about where they came down would be left until tomorrow when we could walk around the yard in daylight to find them.

How we came to never cause an accidental fire anywhere due to our Guy Fawkes celebration remains a mystery, as indeed it does that no one was ever hurt, well seriously. I guess we were just plain lucky.

And then there was the night's finale, which Dad was well prepared for and keenly anticipating. Remember the rusty nail on the door of the powerhouse? That's where he had mounted the one special firework he always bought for the occasion – a Catherine Wheel. A Catherine Wheel was a flat disc about ten centimetres across, made of two separate paper tubes packed with gunpowder, wound around in a spiral from the centre to the outside, each ending in a separate wick. In its centre was a hole, so that you could mount it on something like the rusty nail on the powerhouse door.

Then, in almost total darkness, as the flames of the bonfire had burnt down, we all headed back toward the house, to congregate in front of the powerhouse. With Mum saying: "Cliff? Are you sure this will be alright? Remember you've got all those tins of petrol in there!", and Dad replying that "Of course it will be alright – I've been doing this for years!" He would quickly light both fuses so they both burst into action simultaneously. After a brief period of hissing wicks, the Catherine Wheel took off – in a mad spinning wheel of flame and multicoloured sparks, showering and lighting up the front of the powerhouse. It must have been doing hundreds of revs per minute, slowing down only as it burnt towards its centre and exhausting its supply of gunpowder. What a show to end the night! And somehow, despite all the cracks in the wooden door, and the holes and gaps in the corrugated iron wall, the sparks never managed to ignite any of the petrol just behind the door. The wooden door didn't even catch alight – just a bit charred and blackened.

Formative Years

Guy Fawkes night was just one of the occasions that allowed me to appreciate something that was a particular feature of living where we did. Being able to go out at night and look up at the stars, often free of any clouds, and certainly without any background 'light pollution' due to urban development. The night sky shines bright out at Perenna. Sometimes it seemed so bright and close you felt you could almost reach up and touch the Milky Way

Without the distraction of television, or even after it became available with its grainy black and white image and poor and restricted programming, one of our favourite nighttime entertainments was laying out on the lawn underneath nice warm rugs and looking up at the stars, seeing who would be first to see a shooting star. It was always an awesome and reflective thing to do. I believe all kids need to have this kind of experience if they possibly can.

Then something changed. There was something else up there that drew us out to look for it, with both wonderment and apprehension – Sputnik!

One of the strong influences during my teenage years was the fear of Communism and the so-called 'Cold War': the stand-off between the 'Communist Bloc' led by Soviet Russia, and the democratic Western World. All through my formative years the threat, perceived or real, of nuclear war breaking out between Russia and the rest of the world was very real. I imagine I wasn't alone in feeling very fearful that this may unfold. The stories and recent lived experience of our parents, and other adults, about the recent World War Two played into this. We understood a nuclear war would be far, far worse, most likely bringing about the end of civilization as we knew it. This was the brooding menace of the time. I really was raised in fear of our enemies 'the communists' and the threat they posed, and of nuclear destruction if war again broke out.

Against that background, Sputnik was a really big deal, the first man-made object launched into orbit around the earth – it was a technological marvel, and it was the communists who did it! The threat of communism loomed above us in the night sky! It drew us out to gaze upwards to find this tiny pinpoint of light scuttling across the evening sky. I can remember spotting it, and hearing Dad say – 'Yes, there it is! See how red it looks!' He had me convinced that, of course, a communist satellite would be red – this was the bogeyman of 'reds under the bed' writ large! So began the Space Race which, despite Russia's early lead, culminated in Western dominance with the Moon Landing by the USA not much more than ten years after we first gazed up at Sputnik. The shock of the Russians launching the first satellite certainly galvanised action in America, to whom we looked in our unease. It was with relief that we followed 'our' (meaning our great and powerful friends, the Americans) success in the space race, leading to the ultimate and amazing Moon Landing in 1969.

Of course, our other powerful friends and 'mother' country, the British were getting pretty good at this Atomic Bomb and Rocket stuff as well, and as a kid I was proud to know we (us Australians) were right in there helping them out – by providing half of inland Australia for them to blow up bombs and test fire rockets in – centred at the Woomera Rocket range.

I can remember one night when Dad suggested we should go outside for our stargazing exercise, because "I heard they are going to launch a 'Blue Streak' Rocket from Woomera tonight, and we might be able to see it!" This sounded pretty exciting, so I didn't need any further encouragement – out we went, laid on the lawn, and looked up. But after an hour or so, we agreed the only streaks to see that night was the occasional shooting star or meteorite. It was a bit ambitious to think we would see the "Blue Streak' if it did take off okay, because Woomera is a long way from Perenna, (1500 kms) and they were aiming it in the opposite direction to us.

My understanding is that there were several test flights of rockets at Woomera, but none ever made it to orbit. The only thing that might have floated above us, if we had known about it, would have been radio-active dust from Maralinga where several atomic bombs were set off around the same time.

This was the backdrop to my impressionable teenage years. How should I feel about the future? Pessimistic or optimistic? Was civilisation going to destroy itself, or was there a bright future to look forward to?

Fortunately for me, as I have noted already in this story, optimism won out, so I didn't allow the portents of doom to overburden me, even though some of the leading Hollywood actors of the time, Ava Gardner and Gregory Peck came out to Melbourne to star in a movie called 'On the Beach' – with a narrative of Melbourne being the final major city in the world to succumb to devastating nuclear fall-out. Heady stuff - the end of humanity! But we never bothered to go and watch that movie. Looking forward to Collingwood playing in the finals, or watching the crops develop towards another good harvest was a far better use of our time.

Maybe there was another important aspect of our location. Living out in a relatively flat, open landscape like Perenna might at first glance seem rather boring, but this wasn't necessarily the case, as there were benefits. While at night we had the huge, big and bright starry sky, the daytime feature was the distant horizon. We weren't hemmed in by hills, or the buildings of a town or city scape. Consequently, sunrises and sunsets were often spectacular. It was hard not to be in awe sometimes. Maybe it was the daily ability to look towards the far horizon and imagine what lay beyond as having something to do with an optimistic outlook. Perhaps it was just the hope that the low band of clouds in the west might turn into half an inch of rain. But maybe there was more to it.

I also had the benefit of some good teachers during my high school years at Nhill. Looking back, I do believe that most instilled a positive attitude towards learning, and outlook generally. With one exception, recalling the class discussion with our fifth form Literature teacher who assured us that that from her religious perspective, the Bible's Book of Revelations certainly looked like it foretold our imminent Nuclear Destruction!

I preferred the more interesting upbeat things I was learning about in Science and Geography in particular, and also the strong emphasis on sport, particularly encouraged by the headmaster for most of my six years at Nhill High School, Mr John Bishop.

I don't think it is any bad thing that it was then, and is now, generally considered that sport is an Australian national obsession. And there was no greater highlight of that than the 1956 Olympic Games in Melbourne. I was fortunate that as a ten-year-old this was one of the many occasions that I got to see a big sporting event at the MCG. Sitting in the top deck of the new 'Olympic Stand', (it's since been demolished and replaced by the new Northern Stand,) I was there to witness a couple of key track and field events. We cheered on an Australian success story as Betty Cuthbert won the Women's 100 metres sprint Gold Medal, and then watched the men's 5000 metres final.

Figure 17 A ticket to the 1956 Olympics

This was a different story, with no Australian success – it was won by a communist! I can still recall that event, as Vladimir Kuts from Soviet Russia raced away from the field, leaving all the others far behind. Dad sat there saying to us "Look – the Commo's going to win this by a mile!" They'll never catch him!" Our fear, or disdain of communism, was an undercurrent of the games, as I recall. Of course, it is a fact that in subsequent Olympic years communist nations enjoyed huge success – there were many more to follow Vladimir Kuts.

But there was also lots of success ahead for Australians in many fields of sport, with Test Cricket and Tennis to follow at home and overseas. We spent many a night, sitting by the open fire, listening to the BBC commentary via very scratchy, barely understandable 'Short Wave' radio broadcasts of the Test Cricket in England, and the Wimbledon Tennis. Being able to sit and watch a high resolution flat-screen telecast of these events as we do today wasn't even science fiction in the 1950s.

I do look back on these formative years, and in particular my time at Nhill High School, with good memories.

Getting there then was not much different than today – we caught the school bus, which for most of my time at school started from Netherby and took a circuitous route via Lorquon before heading south for Nhill. We had to be driven four miles down the Perenna Road to 'The Sixteen Mile Fingerboard Corner' to meet the bus. The whole trip took about an hour. The advent of a bitumen road from Nhill to Netherby, which I think was complete by 1958 when I first went to high school, made the school bus service much more comfortable.

At first, going to high school was a bit of a culture shock for me. Nhill High School had an enrolment of over 300, which was rather daunting for a kid who was used to a school with ten or twelve students.

But I must have soon become used to it and enjoyed my years of secondary education. This spanned six years, from form one to the final year of matriculation, which ended my formal education. I never had the benefit of going on to tertiary level – university. So, while I missed out on what I am sure would have been a beneficial experience, I would like to think that my time at Nhill High did leave me with an enquiring mind, and an ability to question things, and to sort fact from fiction.

Sport – the social glue

Life and farming operated at a different pace and intensity in my early years on the farm. We couldn't imagine a time when cropping and harvesting would be sixteen hours a day, sometimes twenty-four hours, seven days per week operation.

Instead, on a typical Tuesday or Thursday afternoon during the footy season, if I was out on the tractor, by three o'clock in the afternoon I would be looking closely at my watch, assessing if I could do another 'round' of the paddock before it was time to knock off, unhook the tractor and drive home so that I would be able to get to footy practice by half past four! It never registered as an issue of great importance that we may not be ready to start sowing wheat until sometime next week – there were higher priorities in life!

The local sporting clubs were indeed central to social life, and all roads led to the Netherby recreation reserve where the footy and cricket oval, plus tennis and netball courts were located. This didn't just apply to Saturdays in the footy season, when if you got there too late, well before the game, all of the car parking spaces on the fence would be taken, and you would have to park back in the second row! But it also applied to training or practice nights, when not just the players turned up. And a host of other social activities were organised around it.

Thus, having parked the tractor in the shed, I would grab my gear and head off to footy practice at Netherby. By the time I got there along with the rest of the team, several other local identities would already be there. Cliff Sanders and Normie Smith would also have parked their tractors, and made their way to the footy sheds where they would have lit up the fire box in the hot water service, which would be well on its way towards venting off steam from a bent copper pipe sticking out of the shed's tin roof, as they fed in more small mallee roots, the standard wood fuel, and cooked up a hundred gallons of water for our showers after training.

Cliff and Norm didn't necessarily have stellar footy careers in their past but were just as keen and committed to the club as anyone or more and would have their white overalls back at home ready for Saturday when they also doubled as club 'trainers'.

Predictably, they would warm themselves by the firebox and be joined by other notables such as the club president and secretary, and of course, the team selection panel, and several others of lesser importance but keen supporters, nonetheless. They also gave footy practice a higher priority than another couple of hours on the tractor, while they waited for us players to take to the field for practice, along with our coach, who would have driven out from Nhill.

The footy wasn't just for the footballers. It was the social glue that held the community together, even training nights. Then the main event - pretty much the whole district would turn out for a Saturday home game.

The men and boys would congregate around the footy shed, while us players would be getting changed and ready for a stirring rev-up by the coach – assuring us that, of course, we could beat these bastards, as long as we gave it all the guts and determination we had! Next door through the thin tin wall you could hear the opposing coach exhorting his charges with the very same message! Then it was time to clear the kids, who had been playing kick to kick out in front of the shed, off the oval, as the players ran out to the sound of a hundred car horns around the ground urging their team on to victory.

Meanwhile, further back around the ground, a team of ladies would have dusted out the afternoon tea shed, emptied out caseloads of sandwiches, cup-cakes and slices onto the benches, and had them all arranged on plates with doilies ready for their clientele, while big kettles were boiled on the stove for tea.

There was a less than salubrious 'lean-to' at the back of the afternoon tea shed where the girls and young women prepared for their netball match – the netball court, which doubled as the tennis court in summer was right behind the tea shed.

Halfway between the afternoon tea shed and the footy shed, two local identities Lloyd Hamlyn and Don McGrice, with a few beers sequestered away to sustain their effort, would be stoking the fire under their old 'copper' with mallee roots, boiling up a heap of 'saveloys' (hot dogs) which with bread rolls and tomato sauce made a tempting alternative to the ladies' offerings. Those hot dogs would boil away all day, and as Lloyd and Don's attention to their condition inexplicably fell away, the few remaining floating around in the boiling water didn't look particularly intact, but at a giveaway price would still satisfy a hungry client.

I am afraid to relate that for all the effort and support, the Netherby Footy team wasn't renowned for its success, both before and after its amalgamation with neighbouring Lorquon, which at the time was considered certain to produce a combined powerhouse of the Lowan Star League. It wasn't to be, with the last premiership glory for Netherby being celebrated some time back in the early 1930s. I'm pretty sure it's correct to say that the netballers were far more successful over the years, but that wasn't rated the same!

But measured by outcomes other than premierships, the Footy Club, and in turn the Cricket and Tennis Clubs over summer made great contributions to community life. They were the social glue that held the district together. It wasn't just that practically the whole district turned out for the day on Saturday. Even in the peak of cropping time, most of the district tractors would be back in the shed by half past ten in the morning, and of course would remain idle on Sundays – farming then was really a nine to five job, five days per week, with little if any work (other than milking the cows) done over the weekend. Both sport and observance of the Sabbath had a much greater importance in family life than today.

Figure 18 Netherby Lorquon FC 1968

I'm not about to make any dubious claims to football prowess on my part, notwithstanding a period as captain of Netherby Lorquon, no less! Suffice to say we were used to absorbing glorious defeats, and when the occasional victory came our way, it was celebrated fittingly. This made the various social events that spun out of the Club's activity all the more memorable. I wish I still had access to the hardcover minute books, and the little case in which they were kept, of the Netherby Lorquon Football Club Social Club, of which I was for some time the secretary. It would be an interesting reminder of those different times.

Now there's another subject altogether – the many local committees, all of which had a full roll call of members from President and Secretary down, and all of which were conducted with the utmost level of correct process and meeting procedure. There was the School Committee, The Fire Brigade Committee, The Hall Committee, The Victorian Wheat and Woolgrowers committee, The Silo Committee, The Recreation Reserve Committee, the Water-Works Trust Committee, The Cricket Club Committee, The Football Club Committee, and the Football Club Social Club Committee. I am sure I have missed several.

And then of course, in those pre-equal opportunity, sexist times, most of those groups had their own ladies committee!

All of these committees required a secretary, the trappings of whose office always came in a small brown or black leather case or briefcase– in which was kept the bound hard cover minute book, plus manilla folders for inward and outward correspondence, copies of the constitution, and other important documents. As is true today, some people seem to gravitate towards the job of secretary, and often end up with many. Dad was one of those, so there was always a neat little stack of secretary's briefcases behind the door in the lounge room. And I must admit, several such briefcases continued to inhabit the house under my subsequent responsibility – including the footy club social club.

Proper process was always observed for the Footy Social Club, notwithstanding the meeting's location at the Netherby Pub. So, for even apparently flippant occasions, all decisions were discussed, and the results entered into the minute book reflecting the formality of process. It would be something like:

> *Moved Vince Dart, seconded Kevin Warner, that we hold a Chop Night on Thursday 19th of June after football training. Carried.*
>
> *Moved John Rethus, seconded Sandy Morrison that we accept Chop Nottle's offer to bring a truckload of mallee stumps. Carried.*
>
> *Moved Chop Nottle, Seconded Vince Dart, that we accept Sandy Morrison's offer to donate and kill three sheep and take them to Gilmour's Butchers in Nhill to be cut up into chops. Carried.*
>
> *(Max Gilmour was a Nhill Football Club legend, plus two of his sons were also stars, so they were the obvious choice of place for this job)*

We would then instruct the catering subcommittee to take over the finer detail of the event – such as the purchase of twenty-four loaves of bread and six bottles of tomato sauce. That should be enough for the three hundred chops that the sheep would be cut into. Do you get the idea why it was called a chop night? And, of course, how much beer was required. Would a 'niner' (a nine-gallon keg) be enough, or should we get an eighteen this time?

All recorded in the minute book. I wonder what happened to that secretary's case after the footy club was wound up?

As you would expect, there would be quite a few more than usual supporters turn up to watch us 'boys' train on the night of the chop night.

While we were doing our 'circle work' and our sprints up and down the oval, they would have unloaded the mallee stumps off Chop Nottle's truck and set up several forty-four-gallon drums cut in half, and got them well alight, with old circular saw blades (they make perfect barbecue hot plates) placed over three of them on which to cook the chops. Then a trestle table for the bread and sauce and beer glasses, with the keg beside. All set to go.

Accordingly, the night would go on, with much discussion while standing around the warming fire drums. The chops were steadily fed onto the saw blades and cooked to varying degrees, which was a bit hard to tell, as we would only have a couple of gas lights, which were mainly required for the beer dispensing area.

Discussion, of course, would start about tactics on how we were going to cause an upset win on Saturday, but then ranged off much more widely. Until, on one particular night, a major problem arose. The beer stopped flowing from the tap!

The problem was soon identified.

"That bloody Lionel Missen!" "The miserable bastard!" "He's given us a half empty gas cylinder!" Lionel was the Netherby Publican at the time. Others would be confirming that there was still plenty of beer in the barrel as they lifted it to check. "There'd be three gallons left in there at least!" Just no gas to force it out. Let down by the publican, who by now would be in bed, as everyone was at the footy ground, and not the pub.

What to do? Typically, ingenuity shone out. Just across the track, parked where the gaslight extended enough to see, was Austin (Ocky) Mackenzie's Commer Truck, with its spare tyre sitting in its rack.

"Hey Ocky" "That spare tyre is pumped up, isn't it?" This was important, as truck tyres are inflated to a high pressure, three times that of a car tyre, and they are much bigger. An ample supply of compressed gas!

In no time at all, Ocky's spare wheel was rolled over, the gas tube ripped unceremoniously off the offending empty cylinder, the valve unscrewed from the truck tyre, and the tube pushed over before much air could escape. It fitted perfectly. Without a leak!

A quick test confirmed success – the beer was flowing again! That left just one small problem, which our apparent thirst overcame. I imagine you have never tried it but take it from me, beer with stale spare rubber tyre air in it does taste different. Very different.

But we were up for it, and the empty barrel, plus the empty gas cylinder with its now inoperative pipe fitting was returned to publican Lionel, the next day. And I never heard if Ocky had a flat tyre on the way home.

I have to relate another chop night held after footy training. This one was at Netherby, because it added somewhat to my developing sense that my generation, and those before me, weren't really conscious of the importance and value of our natural environment.

On this particular occasion, after we had all had had several beers, heaps of chops and bread and sauce, and plenty of talk, it was pretty much time to go home. But there was another option for me – a couple of blokes suggested I might like to go 'Spottin' with them. This was short for 'Spotlighting' – a common practice where you would load up in a Ute after dark – a driver and often a passenger in the front, and three blokes in the back – two with shot guns, and/or rifles, and one with a spotlight, attached to the Ute's battery with alligator clips. The objective was to find and shoot as many rabbits and or foxes as you could. This was, and still is, a legitimate means of pest animal control. The beams of spotlights in the nighttime sky were a common one – it seemed that someone in the district was 'out spottin' just about every other night. I am sure they did have an impact in keeping fox numbers down, and maybe made a bit of a dent in the rabbit population.

But this night was different as I soon found out. Our quarry weren't foxes or rabbits, and our weapons weren't guns. We were heading out through the Netting Fence into the Big Desert in a four wheel drive Land Rover, (this was before Toyota Landcruisers arrived) with several greyhound dogs in the back. We were going out 'Doggin Roos' – that is, to find Kangaroos in the scrub and set a pack of dogs onto them.

It wasn't long into the trip before I started to wonder why we were really doing this. For starters, it seemed to me that the kangaroos we were after were not really a problem – it was rare for us to ever see a kangaroo on our farm, or anyone else's farm in the district for that matter. I realise there are times and places where there is a case for the control of overabundant native animals, but I was 99% certain this was not such a case.

But the main concern I had was that this wasn't sport at all in my view. It was just unnecessary cruelty, both for the few hapless Roos we spotted that night, and, as it turned out, for the dogs as well, one of which was badly wounded by a 'roo as it fought for its life. That night out in the scrub was definitely one of the early experiences that influenced my thinking about our impact on the environment, and why we should respect and conserve it. And if there is a need to control animal numbers, this was not how to do it.

That rather uncomfortable occasion aside, the regular chop nights and occasional pie nights - the bakeries in Nhill made great meat pies - provided the blokes with ample social interaction. But what about the women? It was a sexist world back then, just beginning to change. Thus, if you could read that Footy Club Social Club minute book, you would see that there were many events for both sexes subject to our orderly, formal planning process.

Such as the regular Saturday night dance at the Hall, held on every Saturday night of a home game through the footy season.

True to the spirit of the agreement whereby Netherby and Lorquon amalgamated into one football club, the home game Saturday night dances alternated between the Netherby Hall and the Lorquon Hall. In one aspect, they were different sorts of occasions, inasmuch as Netherby had a pub, and Lorquon didn't!

There were two or three local bands who could provide the necessary music, due to the fact that they were good – they really were – and needed to be booked well in advance. They would provide what was then referred to as 60:40 music – sixty percent rock and modern, forty percent more traditional.

The Hall filled quickly as starting time approached with tickets available in the front foyer. There was, however, a certain well-worn expectation of how the night began. The more senior district ladies would have their cars parked down the side of the hall, for easy access to the kitchen and supper room, where cases, boxes and car-fridge loads of prepared food would be unloaded and taken in ready for 'Supper'. This was ladies' business - I don't remember a man back then ever being involved in the provision of supper – the gender divide was only just beginning to break down.

Already well and truly over-catered, this burgeoning feast was modestly added to by the district's younger ladies and newly-weds, who were expected to bring along a 'plate' of their own. Just a plate of food, more or less as a statement of inclusion, but leave the heavy lifting of preparing the supper to the seasoned campaigners who have been at it for years! It was heavy lifting quite literally – Vera Sanders for instance would turn up with two big old brown suitcases, which had to be carried in on their sides - that was a job for a couple of men - before being opened on the bench to reveal plates full of sandwiches, sausage rolls, cup-cakes and statement-making cream cakes.

In the main hall, the ladies, young and old took up all the seating around the sides of the hall, while the men all stood around at the rear, near the main entrance, awaiting the appearance on stage of the MC (Master of Ceremonies) for the night, who would introduce the band, and invite the gentlemen to choose a partner for the first dance. And so, the night would progress – after each 'bracket' of dances, the gentlemen returned their ladies to their seats, and retreated back to the rear of the hall, while the members of the band repaired side stage for some well-deserved refreshment. Following this format, the evening went on until it was supper time at eleven o'clock, when everyone filed out through the side door into the supper room, to enjoy the feast which was all laid out on long tables along the length of the supper room.

The Saturday night dances were strictly non-alcohol events, which is what differentiated the Netherby and Lorquon dances. With the pub just a short walk from the hall at Netherby, the crowd of men at the rear of the hall was noticeably thinner up until closing time at ten o'clock! Times were changing, but back then no ladies would ever be seen in the main bar, and while there was a lady's lounge at the pub, it wasn't particularly salubrious, so most of the women wisely chose to stay at the hall.

This state of affairs began to be discussed by the Football Club Social Committee, out of which arose the new institution of the Cabaret Ball, which reflected the changing times. This involved the obtaining of a licence to allow the consumption of bring-your-own (BYO) alcohol, and then setting up tables in the supper room with food to be served from the beginning of the night. So, instead of a hall with the women all seated side-by-side around the walls, and the men all standing up the back, everyone all sat down together, and mingled around as the night progressed. Social progress!

This advancement, a true reflection of the rapidly and overdue changing of times and social expectations led to a pressing infrastructure issue. The Netherby and Lorquon halls were no longer 'fit for purpose' (today's jargon!), with their separate main hall and walled off supper room. To allow for free and easy mingling, with the tables in the same room as the dance floor, the walls had to come down. And they did. A combination of local fundraising and government grants saw both Halls remodelled and updated to allow for this new way of a good night out.

The modern age of the 1970s and 80s was about to unfold!

The cricket season was a little different to the footy, as it was not seen as a spectator sport. Unlike the footy, most of the district did not turn up for every game. I wasn't a tennis player, but tennis too was not a spectator drawcard, although it had many participants, men and women, unlike cricket. But cricket was my summer sport of choice, and like the footy, was a priority activity. Although the annual fixture of the Lorquon and District Cricket Association, and its successor the West Wimmera Cricket Association did make some allowance for the importance the grain harvest and the Christmas festive season, and scheduled a 'Harvest Break' over Christmas/New Year.

In keeping with the massive social change that has taken place, the many 'ovals' that once were the focus of football and cricket throughout the district are now long gone, and unless you really knew where to look, you would not recognise these spaces as former sports grounds at all. The main giveaway is the twenty-two yard (20.1metres) long concrete strips that mostly do remain. These concrete slabs were the bases for the cricket pitch, which had a top layer of 'Malthoid', or as time progressed, artificial turf, glued to them.

On the subject of turf, it needs to be pointed out that our local sports ovals had no such thing. The Footy Oval relied on the same seasonal conditions as the district crops – if it rained, a bit of grass and clover would germinate and grow, but for most part, the playing surface was dirt. As the year went on the grass would become more established, providing the footballers didn't grub it all out! By the time the cricket season came around the Lorquon Cricket Oval grass would have dried off, and as summer wore on, the ground would be bone-dry dry grass and dirt. There wasn't even a fence right around the ground, which melded into the nearby crown land reserve, then and now a good example of the original vegetation of the area. The odd lizard, and even a snake, were sometimes trespassers onto the oval, challenging the fieldsman at deep backward square leg or long-off to maintain his concentration on the game.

The only sports ground with real year-round turf was Davis Park in Nhill. We looked forward to our couple of matches per year we were drawn to play there!

Figure 19 LCC - Premiers in 1978/79! – that's me far left

If you drove through the site of the Lorquon township now, it would be very hard for you to believe that here was where my brother Ken and I, and other Lorquon Cricket teammates, would get off the Lorquon School Bus, which had brought us out from the High School in Nhill. We would call into the General Store, located beside the Post Office to buy a 'Choc Wedge' ice-cream and a bottle of Fanta, and walk up to the cricket ground for Thursday Night Cricket practice.

Lamentably, the store, the post office and the cricket ground, indeed the whole township, are now long gone. There is little evidence of their existence, although if you know where to look you will find the twenty-two-yard-long concrete cricket pitch base.

PART 2: THE SEVENTIES AND EIGHTIES

The starting point for this second part of the story of my life at Perenna is a Saturday night in early May 1967, at an MYF (Methodist Youth Fellowship) dance in Horsham, when I looked across the room and saw a shapely, attractive young woman with a big smile. This was the night I met Mary, who became my life's partner and sharer of the rest of my story.

We went out together for some time before our engagement and marriage.

What kinds of things did a young couple do for entertainment in the late 1960s? For one thing, distance was not an issue. Home for Mary was Horsham, one hundred kilometres from Perenna, but travelling that far was no impediment – we all had good cars by the standard of the day - and fuel was cheap, and we were used to driving long distances. It was just an hour's drive to Horsham

That being so, most weekends would involve going to a Saturday night dance or Ball, which were regular events in Horsham. This was the era of Rock n Roll and modern Pop Music – and Horsham had its own home-grown Rock bands, with heaps of new technology amplifiers to fill any space with loud music. If we knew "The Sonamatics' would be playing at the Horsham Town Hall, we would be there – they were our favourites.

An even more likely Saturday night outing would be to the Mota Vu Drive-In movies. This latter form of entertainment was extremely popular at the time. Half an hour before screening time would often see you in a queue of cars stretching for hundreds of metres, a veritable traffic jam, waiting for the ticket boxes at the entrance to process everyone and allow you to drive in and be directed to park in rows facing the big white outdoor screen. Each space had a metal post to which was attached a portable speaker, which you hung inside the car, and settle in to enjoy the movie. The main thing to remember was to replace the speaker on its post before you left, or you would be driving out with a broken speaker cord trailing, and a speaker for which you were liable inside the car!

As our relationship developed, Mary occasionally came up to the farm for a weekend, with one such time standing out - the day she travelled by train to watch a game of cricket at Kiata, where I, of course, was part of the Lorquon team taking on Kiata. Given there is now not even a regular passenger rail service between Melbourne and Horsham, it must be a bit hard to believe that back in 1969 you could buy a ticket on the regular daily train (it was a little rail car called 'The Peanut) that ran from Horsham to Nhill and onwards to Serviceton, and hop off at the Kiata Station, and walk over to the Cricket Ground! In fact, there is very little evidence in Kiata of either the Railway Station or the cricket ground having ever existed – save for a strip of concrete in amongst a heap of native trees and shrubs, where the local community decided the best thing they could do with the disused oval was to plant it back to its original state.

With a shared interest in nature and the environment, we also took many Sunday drives for picnics and walks in the Grampians, a place we continue to enjoy. One of our favourite places was Delleys Dell, a beautiful shady walk that meandered for a couple of hundred metres beneath an archway of tree ferns along the valley floor just upstream of Silverband Falls. We both loved that walk.

The walk along Delleys Dell doesn't exist anymore, its demise almost certainly due to climate change in my mind. The early years of the 2000s were marked by low rainfall, drought and heat wave, which culminated in the massive 'Mt Lubra' wildfire burning many parts of the tinder dry Grampians bare. This was then soon followed up with a massive rainfall event, which dumped over 250 mm of rain over the Grampians in a period of hours. The resulting run off poured off the bare slopes, with landslides occurring right across the Grampians. Many roads and walks were closed for months. In the case of Delleys Dell, the force of water was such that all of the tree ferns, along with the peaty soil in which they grew were swept away, leaving a rock-strewn gully behind. We drive past there now and reflect that we are lucky to count that once beautiful fern gully as a place we had enjoyed.

Generational Change

Figure 20 Our Wedding Day, 18th April 1970

Our marriage in April 1970 coincided with, or more correctly, precipitated a significant change in how the farm was run and managed.

Mary and I moved in as the new occupants of the house at Perenna. In order for this to happen, Dad and Mum (Cliff and Margaret), and our youngest brother Graham, moved to take up residence in Nhill. Preceding this, Des and Shirley had left for further study and pursuit of their careers in Melbourne; to be subsequently followed by Graham, whose career led him overseas. Ken, upon his marriage to Kathleen in 1966, had moved into the house on the 'Clark' property (two kms north) which had been purchased in 1961.

The farm management structure was formed as a partnership between Ken and Kath, and Mary and I, with Dad taking a step back from any direct management.

The issue of how farming businesses change from one generation to another is a complex one, with total fairness to all concerned, both parents and siblings, difficult to achieve without a lot of advanced planning and discussion. The truth is, that little if any planning and discussion took place in our case, with most of the decisions either made by Dad, or just happened or evolved as events unfolded. It's not my intent here to make any judgement or claim as to how well the needs of all were met in our case but do note that in terms of structure and management, it was unusual for the time for us to have the level of autonomy we had to run the farm while still in our twenties.

Looking back, our partnership served us well, but as is always the case, had its limitations and stresses largely due to there being no clear and documented strategy and business plan, and no definition of roles and responsibilities, beyond a year-on-year approach. Decision making was by consensus, which is less than ideal in a modern business setting. As a result, after a couple of decades, it was time to separate and move on, which I am pleased to say we were able to do reasonably amicably.

Mary and I settled in as a new young couple in the district, and the next chapter of my story began. We were the third, and last generation of the Dodds family to live in the house, which we named 'Perenna Lodge', at Pine Flat after Jack and Elizabeth, and Cliff and Margaret, and to raise our family there, with Lisa coming along in 1972, then Russell in 1976, and finally Scott in 1980.

We weren't the only new couple in the district with a young family. The sharp decline in population of our farming district was yet to become really marked, although it was underway. We had the benefit of an active social life involving many other young families who made up a new generation – the 'Baby Boomer' generation. Our social life was less focussed on the immediate small Netherby district, as we became more centred on Nhill as the main focus of business, school, sport, and organisations based in the town.

No description of our first year together at Perenna, and subsequent years, would be complete without talking about mouse plagues. This was due to the fact that while the house was nominally left vacant at the time of our marriage for a period after Dad, Mum and Graham moved into Nhill, and we honeymooned touring NSW, and cruising on the Hawkesbury River. Actually, it wasn't vacant. It was occupied by a horde of mice! Mary had quite an introduction to living at Perenna – a welcoming committee of dozens, if not hundreds of mice scurrying throughout the house. To say nothing of those over at the barn, where they were in their thousands! Consequently, our first few weeks involved lots of setting of mousetraps, plugging holes in the walls with steel wool and 'Spackle', and waking the cat up whenever he decided he was too tired from chasing and catching them. We slowly asserted our authority as the rightful occupants. And to enjoy living without the reeking stink of mice!

It wasn't the only mouse plague we had at Perenna, or indeed right across the Wimmera Mallee. From a mouse's viewpoint, having lots of farmers around the place, growing all these grain crops, and leaving a percentage of it lying on the ground after harvest, or as spillage around the farmyard, was a very good idea. They had an ample food resource, and plenty of shelter in the stubbles, in and around all the farm buildings, in haystacks, and under loose sheets of corrugated iron, etc. All they needed was a bit of mild weather at the right time to breed up in large numbers, and bingo - millions of mice! This seems to happen every few years, and I believe more often now than in the past, as the move to zero till farming has provided less disturbance, and more shelter for them.

And mice are not just a problem in the house. When in plague numbers, they can have a big impact on crop production. As years went on, we certainly witnessed this. They can cause big damage to emerging crops, soon after seeding. As soon as the tiny shoots begin to appear, they take this as their cue that a nice seed is buried shallowly there, and dig down to get it, leaving your paddock full of drill lines pockmarked with thousands of small mouse holes, with a dying shoot left lying sideways, its nourishing seed and tiny new roots eaten and gone. The worst example of this that I recall all too well was in the early Nineties, when we had purchased a truckload of expensive seed for a new Chickpea variety named 'Lasseter' (How could we go wrong?) and seeded it into what seemed to be perfect conditions with good soil moisture. A good germination was assured, but we had not allowed for the numbers of mice that were around, and how we were playing into their hands by our new more accurate method of Direct Drilling. This placed all these nice new big seeds into the bottom of nice drill lines all thirty centimetres apart, with only a couple of centimetres of covering soil. A perfect way to seed a crop, but also perfect for the mice. All they needed to do was to follow the drill lines, almost as if we had drawn a diagram across the whole paddock for them. Over a period of what might only have been a week, the newly emerging crop of chickpeas was almost completely destroyed. It wasn't just the chickpeas they attacked that year, but the damage to that large-seeded crop was the most severe.

Of course, we weren't the only farm where mice were benefitting from the widespread adoption of Zero Tillage and Stubble Retention practices, leading after that mouse plague to the widespread adoption of baiting with rodent poison at or soon after seeding time, to reduce such damage and loss when mice numbers are high. This was one of the many trade-offs that have been required in order that the undoubted environmental benefits of reduced tillage can be realised.

The Hot Water Service

While the overriding theme of my story is one of big social change at a district level, I do need to recognise the change that Mary experienced as a young woman coming into this farming way of life from the outside. As it was for others like her, because there were many new couples like us, and most of the new young wives tended not to be daughters of farming families, but newcomers to the district. That was an ironic thing in itself – most of the girls from farming families really had no choice but to leave and further their education or pursuit of work or career elsewhere, as it was never considered then that they might actually become farmers themselves.

Mary's adjustment to life on the farm and the district no doubt had many aspects, but for now I will focus on a couple – the Hot Water Service, and the Tin Kettling

The Hot Water Service at Perenna was a something that hadn't changed for a very long time – at the very least from my earliest childhood. It was actually a wood-heated furnace, located in a corrugated iron 'Lean-To' at the rear of the house. It was a rectangular thing made of steel, kind of like a stove, with an enclosed fire box with a grate and ashtray beneath it. It had a chute with a hinged metal lid, where you loaded and reloaded the small split pieces of firewood, or brown coal 'briquettes' (made in the Latrobe Valley, the other side of Melbourne) you could buy in Nhill, and a metal flue at the top, leading up through the corrugated iron roof. It had a metal water tank above the fire box, surrounding the flue, which is where the water was heated. And sticking out of the top, through the roof beside the flue, was copper overflow pipe which hissed with steam once you got the water up to boiling point. No such thing as a thermostat control to make sure the water never got above 60 degrees for safety on this device! The finishing touch to this unit was its cover of insulation. Its rear, sides, and top were all clad in a mix of Plaster of Paris and straw, providing a nice soft, white (becoming very grubby over time) exterior to the whole thing. How or why the straw never caught fire I'll never know! Now a really good aspect of this soft covering, particularly on the top, was that it was the ideal place for the pet cat to sit and keep himself warm, snuggled between the flue and the overflow pipe. A perfect place on a cold winter's morning!

The next part of this story was that soon after we got home from our honeymoon, cruising on the Hawkesbury River near Sydney, it was time to work the fallow and plant the crop, so I was pretty busy out on the tractor. Now Dad (CT as we referred to him) chose one morning while I was busy to come out from Nhill with the purpose to instruct Mary in two very important things:

1. *How to ensure that I always ate a good big steak on Friday Night before football*

2. *How to light the Hot Water Service*

The first was accomplished by him showing Mary the big paper wrapped parcel of steak he had bought at Gilmour's butchers and placing it at the front of a shelf in the fridge for her, making sure it couldn't be missed. None of these new-fangled

1970s diets when a good performance at football was at risk. I reflect, not without a slight touch of irony, that today's Sports Nutritionists would be far more likely to endorse the pasta that Mary was likely to have favoured on a Friday night rather than a big slab of red meat!

He then went on to a demonstration of how to light the Hot Water Service. The first step he showed her was to get the small jam tin with its top removed, which was kept on the top of the hot water service, on the opposite side of the flue to the cat. You took this little tin over to the big shed where there was a forty-four-gallon drum of kerosene with a pump in it, filled it up and brought it back, which he had already done. You then got a few sheets of newspaper, crumpled them up, and stuffed them down the chute into the firebox. Then came a few handfuls of wood chips or kindling, followed by several pieces of nicely split firewood, which was stacked against the tin wall of the lean-to, and a few briquettes, made from Latrobe Valley brown coal. You then took the tin of kerosene and poured it down the chute. This to be followed by striking a match, tossing it down on top of all this, and quickly flipping the metal lid shut. And then you stood back.

The resultant rather rapid starting of the fire generally involved the metal lid popping up and then clanging down to shut itself again, a bit of a creaking of metal joints in the corrugated iron roof, as the flue shook, enough to cause the cat to briefly raise an eyelid, before returning to his slumber. Half an hour later there was steam puffing out of the overflow pipe, and hot water for the day was assured!

What followed a couple of days later was apparently a bit more dramatic – I recall it well. I came back to the house around lunch time, to find Mary a little dishevelled, and very eager to explain along the lines that the bloody hot water service had blown up and nearly killed her, and she had no idea where the cat was.

She went on to explain that needing to heat more water for the washing, she had followed CT's instructions diligently, but with one small problem, being that when she went over to the shed with the little kerosene tin, she found two forty-four-gallon drums with pumps in them, not one. This hadn't been explained. After a bit of consideration of which drum to use, she unfortunately made the wrong choice and came back to the house with a tin of petrol.

I think the next part of this story is self-explanatory. It was fortunate indeed that somehow Mary escaped with a huge fright and no other injury – she must have been cautious enough to quickly stand well back after she threw the match in. Petrol being multiple times more volatile than Kerosene, it would have been a truly serious explosion that started the fire, which could be seen by the fact that the iron roof of the lean-to was relocated about six inches from its original position.

I'm sure that this experience was a contributing factor to Mary's well-known dislike to being anywhere near a gas barbecue when it is being lit. And as a further postscript, the cat eventually did turn up again, late that night.

And given we did have the benefit of 240-volt mains power, Mary and I did sometime later decide that a modern Electric Hot water Service would be a better, safer form of water heating.

The Tin Kettling

It wasn't long after the Hot Water Service incident, we found ourselves exposed to a different kind of happening, this time a long-standing ritual of welcoming newly-weds in country areas. It was about eight o'clock in the evening, and we were in the living room in front of the fire, nice and relaxed, with the cat also stretched out warming himself - he had regained his trust in us and our ability to light fires without too much drama - when a whole lot of metallic sounding noise just outside, and seemingly all around the house drowned out the TV.

We soon realised that this was something we were half expecting – the old country tradition of a Tin Kettling! This involved a whole lot of neighbours and others from throughout the district choosing a particular night, but not letting the newlyweds know, to descend upon the residence, and equipped with an array of metal objects – saucepans, tin cans, metal buckets, whatever would make a heap of noise when clanged together or hit with a stick, augmented greatly if there was a corrugated iron fence around the house, as there was in our case, which could be enlisted to make more noise.

Tradition has it that this noise can only be brought to a stop by the newlyweds coming to the door, expressing surprise, and inviting everyone in for a 'meet and greet' session. In accepting this invitation, the throng quickly produced an array of food and drinks all prepared for the occasion, so as not to place a burden of hospitality on the new couple.

Now as it happened, this latter planning for food and drinks for our 'tin kettling' was taken to a different level, largely for the benefit of all the menfolk and for the questionable benefit of the groom.

Part of all the noise had been 'Chop' Nottle's Bedford Truck being driven into the yard over on the far side of the machinery shed away from the house, followed by a couple of Utes. The truck had a load of three or four tonnes of Mallee Roots (Stumps), and the Utes had trestle tables, some forty-four-gallon drums cut in half, some old circular sawblades, plus an eighteen-gallon keg of beer.

The stumps were tipped off in a big heap, and with the help of a bit of petrol were soon well alight, with some also transferred into the forty-four-gallon drum bases, with the saw blades placed on top as BBQ hot plates. The trestles were set up in the shed to hold the required beer glasses, loaves of bread, bottles of tomato sauce and chops and sausages.

So even though the year was now 1970, the developing movement towards equality of the sexes, and social co-mingling was not on full display the night of our Tin Kettling at Perenna! The blokes were set for a good night! Due deference to the women folk was paid first, particularly as this was the celebration of a recent marriage, with a notable and more senior local gentleman - I think it was 'Sandy'

Morrison – actually a good fun-loving bloke and friend and sporting mentor of mine, appointed to make a nice speech of welcome and good wishes to the new couple. Then, Mary found herself to be the hostess for half the district's women, with a few bottles of wine, lots of lemonade and heaps of chips and cakes and slices, while I was invited to join the blokes over at the shed. There we stayed, until the keg was emptied, and the three tonnes of mallee roots were burnt down to a pile of ashes.

I do recall that Mary and some of the younger women did venture over to the shed for a while but decided that they didn't want to talk about tractors and football all night, so repaired back to the warmth of the house, also provided by the burning of mallee roots in the fireplace.

I'm not sure if Tin-Kettlings have receded into history in rural communities since 1970, but I cannot remember ever attending another one after that night of ours. Maybe the propensity, as demonstrated, for them to divide into a women-in-one room and men-in-another (or over at the shed) affair no longer fitted with the changing social times, and the expectations of younger women. Or maybe it was that suddenly having a hundred uninvited guests was no longer considered a reasonable imposition on newlywed couples. In some ways that would be a pity because spontaneous social interaction can be both fun and beneficial, particularly as social isolation is increasingly a societal problem, magnified, I believe, in rural districts as population decline has accelerated.

Time for New Ideas

Thus, as the next generation to farm the land out at Perenna, we set off into a new period of changing times. With respect to how the farm was managed, there had to be a stronger emphasis on the farm being seen as a business rather than just a way of life. This is not to belittle what had gone on before, rather a reflection that it was becoming increasingly difficult to maintain the standard of living we had grown up enjoying without a stronger focus on the actual business we were in, and how to make it, and keep it increasingly profitable.

I guess I don't really know if it was pretty much universal, but I suspect the business management processes we inherited were common for the time – you worked hard and grew your crops, and produced your wool, sold a few sheep, and had money in the bank. You kept your bank statements, cheque book stubs, payment advices and statements from the Wheat Board, the Barley Board and your Stock and Station Agent in a leather brief case (or a shoe box!), to be plonked on the desk of your solicitor once a year when it was time to 'Do your Tax'. It was his job to produce a statement of Income and Expenses in order that an income tax return could be submitted.

As a result, one of the very first things that became apparent as we took responsibility for the business, was that it was time to set up a proper financial recording system, so that you didn't have just this one point in the year when you knew if you were making a profit or not! The benefit of having studied the basics of 'Commercial Principles and Practice' at Nhill High was a bit of a help, as was Mary's input having come from an office administration role in Horsham.

There was good timing here in that our takeover of managing the farm pretty much coincided with the arrival in Nhill of Hugh Wynter, a Farm Management Advisor. His establishment in the area as a Private Consultant was the result of an initiative of the Nhill Rotary Club, of which Dad (CT) was an inaugural member. Hugh's background was originally from England, having arrived in Nhill via New Zealand where he studied Agriculture at Lincoln near Christchurch.

I well remember the first meeting with Hugh, as a result of a phone call from Dad – suggesting we might like to have a visit from him, as

"I've heard he's a pretty good Bloke. He gave a talk to Rotary last night, and he might have a few ideas".

It turned out that Hugh did have a few ideas – that was, after all, his role, to assess current farm management practice in the district, and suggest areas for improvement or change, for those who chose to engage him, to provide this independent source of information and advice. Looking back, his arrival in the Wimmera did represent one of the significant changes that was about to take place – farmers moving from an era where the knowledge base was largely from father to son, with limited (free) access to government agency (Department of Agriculture) staff in their role as 'Extension Officers'. The new era, which I think we were fortunate to be part of, was a new layer of professionals available to help build and maintain farm businesses - accountants, agronomists, financial and farm management advisors. All important factors as farming became both a business and a way of life.

After that first meeting, we did engage Hugh as our farm advisor, and he was influential in how we established ourselves as a business, and the learning that went with it. At a personal level, I enjoyed having such an "Influencer" as I developed my thinking.

Dad called up to ask "How'd you go with that Wynter bloke?" to which I replied along the lines of "Pretty good. He reckons we should employ an accountant, start doing Cash Flow Budgets and use a Multi Column Cash Book to track everything, and we should double the rate of superphosphate we use when we plant the crop."

There was a slight pause on the phone, as the information that a fair bit of change might be afoot, putting previous practice under question sunk in, before he replied:

"You don't want to take too much notice of him. After all, what would a Pommie know about farming in the Wimmera?"

To be fair, Dad had stood aside and gave us free rein, and followed us with interest, and I think with some pride as we progressed, even if we did take advice from such sources.

As a result, with Hugh's guidance we did start preparing Cash Flow Budgets, and set up a Multi Column Cash Book, and we increased the rate of Superphosphate we used, among other things.

A key reflection with respect to the budgeting and bookkeeping system we set up was how it quickly led us into another truly new time of technological development – the invention of microprocessors, that led first to handheld calculators, and then computers. If you might ever have maintained a cash book with thirty separate columns of figures entered depending on the nature of the expenditure or income item, you will know the level of concentration required as you added up all the numbers in every column in dollars and cents at the bottom of each page, and transferred them into the carried forward line at the top of the next page, and to produce monthly sub totals, then leading to final end of year figures. Quite a task, underscoring the importance of all that time you spent at school studying arithmetic!

Until the invention and availability of calculators. How amazing it was to have our first such thing, just adding the numbers as simply as it was to enter them on the keyboard. What a labour-saving device that was!

The use of a calculator was amazing enough, but the next development was even more far reaching.

This arrived via the day, probably in the late seventies, when Hugh Wynter came to help prepare a cash flow budget. He came into the house with his brief case and another large case and plonked it on the kitchen table. He opened it up to reveal the first portable computer I ever saw. This was the day I was introduced to electronic spread sheets, allowing you to freely change figures for future income and expenditure as you modelled an annual budget. What a breakthrough, replacing the need for endless paper-based revisions and rewrites (and use of white-out fluid) as you tested scenarios for the year ahead.

It was some years before we did acquire a computer (or computers) of our own in 1983, as Wynter's Unit was clearly very expensive, and could only be justified by his using it to help budgeting for all of his clientele. But that day was the day that set in train my, and our use of computers, as a key tool of business management. And what clunky, slow things they were, looking back, but at the time they were extraordinary new tools. What a long way we have come since those machines, which when started came up with a blank black screen with C:\> in the top left-hand corner! The first computer we purchased was a "Commodore 64" – and I well recall the hours I spent figuring out the first rudimentary spreadsheet and word processor programs it had. And attending a (very) basic computer programming course at Longerenong College.

A final aspect of Wynter's work for us was, in a kind of circuitous way, an influence on my thinking about the natural environment. This led from a discussion around the kitchen table with him about whether we should look into 'developing a scrub block'. There were still several farmers in the district who were working on clearing the Mallee Scrub on the edge of the Big Desert, for the purpose of mainly grazing livestock. Hence, we discussed if we should join in, suggesting an example of 4000 acres that might be available just north of the Netting Fence.

This was before he had his computer with an ability to do a "what if" analysis on the spot, so he said: "Leave it with me, and I'll do some numbers for you."

A few weeks later he was back with a heap of notes, where he had researched the cost of clearing the scrub, fencing and water points, pasture development, cost of livestock, and the ongoing costs of fertiliser, set against the present and predicted wool and lamb prices. His answer was fairly unequivocal: "You'd be lucky to make any money at all on that country out there with its low rainfall and poor soil. You might even lose money. You would be much better off to lease or buy some of the much better land around and south of you."

They were wise words indeed, and we did follow his advice. We didn't get into the scrub clearing business.

This all took place at or towards the end of the time when there was much controversy about proposals to develop a large portion of the Little Desert south of Nhill for farming. There were strong proponents, particularly businesses in Nhill and elsewhere who could see business growth as suppliers to all this activity, and they had friends in high places in the state government including then Minister for Lands, Sir William McDonald. But there was also strong opposition from both people arguing that the new farms would be un-economic, (I'm not sure if Wynter was one of these) and from the growing Environmental Movement, both local and state-wide, arguing that the area should be preserved and declared a National Park. The latter cause won the day, and the creation of the Little Desert National Park is generally credited with being a key milestone in the preservation of Victoria's biodiversity. Subsequently a very large area of the Big Desert was included in a much-expanded Wyperfield National Park, although this doesn't include the area just north of the Netting Fence at Perenna that we were interested in.

While I was not actively involved in the politics of the time of land clearing versus the environment, I am very pleased that further clearing of those areas did not proceed, and it was preserved as a vital refuge for our native plants and animals, and for a place for us to visit and appreciate. For it to have been lost and replaced by uneconomic or barely economic farms would have been a travesty.

I'm proud of how we did emphasise the importance of the farm being a business to be well managed and not just a way of life, because your way of life would be much less to be celebrated without a viable business underpinning it.

That said, it's my reflection that Mary and I, and Lisa, Russell and Scott did have a lifestyle out there at Perenna that did have things going for us that we may not have had elsewhere, and we did take advantage of.

Like the twenty-acre timber paddock that was essentially our playground, particularly as we destocked it so that it was without any house cows or other grazing animals and allowed to regenerate as a natural area. Therefore, without anything to concern us, other than perhaps the odd snake, the kids were able to run free outside the house for quite a distance. Be it on foot, on bikes, or, in later years, in a Go Kart. We had a network of tracks or pathways, even a flying fox and picnic ground.

Like being able to learn to drive a car or Ute or ride-on mower as soon as your foot could reach the pedals.

And like being all involved in the farming operation, as we went from seasonal task to seasonal task.

Drought and Floods

The intent of this tale is not to fall into the trap that so characterises stories of rural Australia – archetypal battlers fighting an endless struggle against the elements and authority to eke a courageous but sparse living from the land. I would far more hope that my recollection is seen as one of generally good fortune, where we took on the challenge of having to adjust and adapt to the seasonal conditions that apply in this driest of continents and were successful. We enjoyed many good seasons – and some of my very best memories are of watching good crops develop with enough rain to produce average or above average results, and the great feeling of watching grain from a bumper harvest pour in at Harvest time. We enjoyed many good seasons and good times.

But I guess it is part of our makeup that events falling into the category of being 'Bad' do take up a special place in our memories, long to be recalled.

There were two bad drought years that loom large in my memory on the farm – 1967, and again in 1982. There were other years of significantly lower than average rainfall, 1972 in particular, but these were the two years where rainfall was so low

Figure 21 Checking a Wheat Crop -1995

that most of our crops failed – little if any yield. As farm disasters go, droughts are a special case, in that they don't just suddenly happen, but take months and months to slowly unfold. You cling onto hope that rain will come before it's too late, and it's only right at the end of the season when the crops finally shrivel and die, that you realise that it's all over – it's a serious drought. That's pretty much what happened in 67 and 82. There were some pretty bad years of low rainfall in the early 2000s - the so-called 'Millenium Drought', but Mary and I no longer relied on the business of farming by then, so were only marginally affected – our main disaster being our garden at 'Mooringa' withering away due to lack of water.

The 1967 drought did have a big impact, because it was the first really dry year since the war – before my time. The seasons had been consistently good right through the fifties and sixties as discussed before, perhaps lulling us into a false sense of security that the 'bad old days' of depression and drought through the 1930s and forties were unlikely to return. I do recall Dad, in his typically optimistic way discussing with us, and the neighbours, how since we had moved to modern ways with tractors instead of horses, because of which we did everything so much better, we were pretty much 'Drought Proof'. A version of the 'Rain follows the Plow' mindset that had led farmers to settle beyond the famous 'Goyder Line' in South Australia decades earlier? And most locals, with the exception of a few pessimists or old wise heads who counselled caution, agreed. Thus, having widespread crop failure of 1967 was a jolt back to reality for many of us.

Among key recollections of the '67 drought was how hot and dry the summer was, and how dispiriting it was to try and harvest what little grain we could. This involved driving the tractor and header round and round a paddock for hours without filling the grain tank, dragging the header 'comb front' along the ground trying to gather the few heads that had formed just a few inches high. This resulted in sticks and stones and clods of soil all being fed into the machine, leading to much greater wear and tear and breakdowns than normal. It was a real achievement "Just to get our seed back", so that we, at least, would have enough grain stored to enable us to plant the next year's crop.

And the heat and dry. Still stuck in my mind was one particular day when we noted the temperature at 115 degrees Fahrenheit, (460C) with a searing, dust-bearing northwest wind. I looked out over the corrugated tin fence that surrounded the house, and could see nothing but bare, hot dirt, and dust. I resolved there and then that we needed more trees for shelter from such conditions if this was to happen again. I followed through on this the next autumn, driving down to the Forest Commission nursery at Wail, and coming home with a box of Aleppo Pine seedlings, which we planted in a row just outside that fence down the west side of the house. They grew, and those trees did provide shelter for us in the hot dry times to come. I think that was my lightbulb moment which led to my subsequent interest in the need to plant more farm trees.

And after Mary joined me there three years later, we had a shared interest in making 'Perenna Lodge' as we named it, a much more sheltered and greener place, a retreat from the big wide-open acreage surrounding it. That tall, dense row of Aleppo's really formed the basis of our little 'Oasis' as we planted our way to connecting what was an open windswept house yard into our timber paddock. We removed the tin fence, planted more native trees and shrubs, and extended the lawn area, leading out into 'The Timber'

The best part about a drought, is how good it is when it finally breaks. The land bursts into life again, often to an amazing degree. Many of the very best years we experienced were those that immediately followed a drought. It was observable that crop and pasture growth was vigorous and healthy beyond 'normal'. This was not just because good rains had returned – there were other factors at play also. I think it is fairly well understood now that long periods of little or no plant growth are nature's way of fallowing or resting the ground, in particular by the reduction level of soil pathogens such as fungal diseases and root nematodes.

Accordingly, just as those 'bad' drought years are fixed in my memory, so too are the good ones that immediately followed. Like 1968, when we, and pretty much every farmer in southern Australia grew bumper crops, with record yields, with vast amounts of grain delivered to handling authorities for storage and subsequent marketing. We had beautiful crops of 'Pinnacle', 'Insignia' and 'Olympic' varieties of wheat and worked long into January to harvest it all. The silos at Netherby were stuffed to the rafters, while we waited for trains to come and take some away to the 'Central Receival' storage at Murtoa, or further on to Geelong, so we could resume delivery.

Paradoxically, Australia's record crop in 1968 coincided with high grain production overseas, with a resultant slump in export demand. This surplus in supply led to one of the first big challenges for our subsidised 'orderly' marketing arrangements, leading to the imposition of wheat 'Quotas' by the authorities the following year. This meant that the government guaranteed (Australian Wheat Board) price would only be available for a certain percentage of your recent past wheat deliveries. As there was essentially no other market outlet than the AWB, you had really no choice but to try to grow enough for your quota and little more. This was a highly controversial intervention, causing all sorts of claims of unfairly calculated quotas, and really threw into question how the government backed orderly marketing of grain, which had been in place since the war with strong grower support, should continue. As a result, we were only able to grow 65 acres of wheat in 1969, down from our previous 200-300 acres.

Governments of both sides of politics grappled with how to structure 'orderly marketing' in the years that followed, with much debate and division within the industry. The end point was the deregulation of grain marketing which was pretty much complete by the early 1990s. The institutions and arrangements, so strongly favoured and fought for by Cliff Dodds and most of his peers back in the fifties, were no more.

The year following the 1972 drought was another bumper year of lovely, healthy crops and pastures, again demonstrating a post drought response. In this case it wasn't just good wheat crops - the memorable bit being the clover pastures we grew, and what happened next.

With an early autumn break in 1973, and then continued follow up rain, the district soon changed from brown to green. We didn't know it at the time, because I don't think the Weather Bureau had invented it yet but were we at the start of one of the big 'La Nina' weather events that had major effect across Australia. We certainly didn't have any computers to look up the Bureau's Outlook, and if there were such a thing as satellite images of the weather, we wouldn't be able to make much of it on our grainy black and white TV anyway. Therefore, we relied on observing clouds of insects over the Mallee trees, looking at 'waterdogs' in the sky, and what the ants were doing, to make our own predictions about the chance of rain. All these indicators suggested the drought was over, and it was going to rain. But we didn't realise how much.

With warm moist soil springing back to life after a year or more of drought, the paddocks we had left out from our crop rotation soon had a wonderful cover of 'Medic' clover. As we had destocked the farm during the drought, we only had a small flock of breeding ewes to graze it. So while we drove the tractor round and round our cropping paddocks, continually cultivating out the new weed growth after each successive and repeated rain event, before finally hitching up the 'combine' Seeder and drilling in the seed and superphosphate, the clover just grew and grew. So did the crops when they came up in late June, but the clover had a head start of two or three months. And it kept on raining, and the crops and clover kept growing. And water started to run around the district, gathering in the many local swamps or wetlands. Notably, the Wimmera River ran for months at times exceeding flood level, with the result that Lake Hindmarsh which had been dry since the 1967 drought, was well on its way to filling.

As spring approached, it was obvious that the sheep were never going to eat all the clover pasture, so we decided it would be a good plan to 'lock up' three hundred acres of the best clover, and get into the business of producing hay for the market – a market we had never previously been in. We had the prospect of a big crop of high-quality clover hay, more than we would ever want for our own livestock in the year ahead. All we had to do was cut it, bale it, safely stack it up, then sell it, keeping a bit of it for our own future use.

As we were looking at thousands of bales of hay, our existing hay handling capacity was probably going to be overstretched. That was to walk along beside the truck as it was driven along, pick up each bale by hand, and throw it up to the stacker who built a load that wouldn't fall off on the way out the gate and down the road to the haystack site back at the farmyard. Not exactly 'state of the art'.

Getting it cut and baled was okay, as our neighbour Sandy Morrison did contract hay baling, but how to move and stack it was the issue. The answer came through

Hugh Wynter's Farm Advisory Network, as one of his clients at Yanac had a similar issue, and he proposed that a syndicated plan to buy and operate a newly available 'Stack Hand' machine would be a justified expense shared across two properties. As a result, we ordered one of these rare new machines, which was operated by towing it along with your tractor and just watching in wonder as it picked up the bales and stacked them on its hydraulically operated trailer until you had the equivalent of a decent truckload. When it was full, off you went to a stack site, reversed up, operated the hydraulics and a little stack of hay was neatly released, standing up like a Stonehenge pillar. Each load was backed up to the last, and you slowly built a neat long haystack, without any manual labour. What a machine!

We cut the hay, and Sandy came and baled it all up – thousands and thousands of bales of top-quality clover hay. When it was cured and ready to be brought in out of the paddock, it was our turn first with the stacker machine.

Prior to starting, we identified a new haystack site in the Timber paddock and graded it nice and flat, and began bringing in load after load, and stacking it all together. The haystack grew and grew – a nice long narrow edifice about four metres high, and probably a hundred metres long. At this point, Dad came out from Nhill, keen to see this amazing labour-saving hay carting machine in action. He was suitably impressed but did in passing point out that he hoped we didn't get a flood like the one in 1930, because back then the site of our magnificent haystack was probably four or five feet deep in water! I can recall thanking about that a little but put it down as another of his exaggerated stories, and that something like that would never happen again. Hadn't we just had a really wet winter, and there was no water anywhere near the site? That was all over in the low area in the middle of the Timber paddock - that's where the water would have been in 1930!

However, given that the stack was already big, and we had only carted half the hay in by then, we started a new haystack up on the Perenna Hill – and it grew also to an impressive length. All that was then required was to purchase several rolls of the newly available 'Blue Polyweave' light tarpaulin material, and rolled it out to cover the stacks, securing it with lots of ropes and bale clips, leaving it neatly protected from further rain. Until they all blew off on the next strong south-westerly wind, forcing us to unravel the tangled heap of tarps, and putting them all back on again, tying them down better this time. Until they blew off again. I think that after one more futile cycle of that exercise we decided it was summertime now. It wasn't likely to rain any more, and we were too busy on the wheat harvest anyway. And it was another bumper crop – we were busy. The half-shredded tangled up tarps stayed off.

And it didn't rain much at all through most of the harvest, which produced one of our best year's crop results ever, despite a spring outbreak of stem rust. What a year it had been, following on as it did from a drought.

As the seventy-year record of rainfall at Perenna shows (See Appendix), 1973 was the wettest year ever, with 663 mm recorded, almost twice the 367 mm average.

But the La Nina wasn't finished yet – not by a long way, as 1974 unfolded. And the highlight was the record fall of 192 mm which fell in a single day – Sunday 27 January 1974.

During the week preceding, all our wet weather indicators – clouds of flying insects hovering over the mallee trees, the ants building up their nests etc., were in overdrive, as was our own observation of tropical heat and humidity with billowing clouds developing each day, culminating on the Saturday as a massive cloud bank approached from the north, creating an almost eerie green darkness. Just the night to go visiting our friends Roger and Joan Gordon who lived 40 kms away at Broughton!

We had barely arrived at their place when steady, heavy rain began, with lots of thunder and lightning. It was really quite exciting – Roger and I played billiards while enjoying a few beers as the night wore on, and the rain teemed down outside. Every half hour or so Roger would run out across the back lawn to the rain gauge on the fence, well illuminated by the incessant lightning, and report back – "We've nearly had an Inch!", and then soon after "It's an inch and a half now!". I think it was just after midnight when he reported "two and a half inches!"

At this point, we made a rather silly decision, and decided we had better go home. We tucked Lisa, who was just over twelve months old into her little bassinette, making sure the cotton safety net fitted over and was secured by its elastic band – the most recently available safety feature for carrying babies in cars, and placed her on the back seat. And off we drove into the night, with thunder, lightning and teeming rain, and water everywhere.

It wasn't far at all down the gravel road towards Yanac that it became obvious that the drive home wasn't going to be easy, or without hazard, or very sensible! All the roadside drains were full of water and overflowing, with long sections of the road itself underwater. We quickly decided that driving home via the Yanac/ Netherby Road, which was a gravel unsealed road, with several low 'inverts' where the water was likely to be deep would be too risky, so we turned south on the sealed road to Nhill. But even the sealed road, which was built up with culverts to allow water to cross under the road turned out to be overwhelmed by all the water. There was a large section just south of Yanac where water was pouring across the road, headed for the nearby Yanac Swamp. Water must have been at least at door level as we negotiated this, and we were both sure the car actually drifted sideways at least once. And this was repeated again as we encountered more flooded sections. We made it to Nhill, with the rain still pouring down, but decided if we had made it that far, we should be able to drive out from Nhill to home. And surely the rain must be going to stop soon! On we pressed, with water over the road in many sections, again with the car feeling pulled sideways as the water flowed towards the local swamps. But at least the road was solid under us, until we approached what we called the 'Fourteen Mile Fingerboard'.

It was here that we hit a deep 'wash-away', where the bitumen surface was apparently gone, with the car bouncing heavily through deep water before landing back on a solid surface. This launched Lisa's bassinet off the back seat into the foot-well, with Mary quickly reaching back to retrieve and settle her. That little net must have worked, as she was still safely inside, and okay. The car seemed to have no apparent damage, and the engine was still running, but this was where, from my point of view, the trip had become decidedly scary. Fortunately, the last ten kilometres was the least flooded, and finally driving into the gate at home was with a great sense of relief. I think it was between three and four o'clock in the morning, but we were home safe. And still the rain poured down. That late night drive was through the heaviest downpour of rain ever recorded in the Nhill District.

We fell into bed. I think we slept for a couple of hours, before waking soon after daybreak, realising the rain had just finally stopped. This really was a truly significant weather event, and as such I retain vivid memories of it and the aftermath.

The first was measuring the rain. I went out to the fence post where the rain-gauge was mounted and saw that the measuring glass which held enough for one hundred points or an inch of rain on the old scale (25mm) had well and truly overflowed into its metal container. I then brought the whole thing inside and placed it on the kitchen table while Mary got out a big glass water jug, and a notepad and pen

We tipped the gauge's content into the water jug. And then began to carefully pour it into the measuring glass, filling it to the one-hundred-point mark, noting it on the writing pad, before emptying it down the sink, and measuring the next inch. Mary had seven inches recorded before the last bit in the jug only made it halfway up the glass. The figure was an unprecedented daily record of 757 points of rain! – 192 mm

The result was district wide flash flooding, with water draining into all of the lower lying swamps or wetlands and covering large areas of farmland. As I went out to the rain gauge, the first thing I noted was that the timber paddock or 'Pine Flat' had become a small lake, with water well beyond anything I had seen before, probably at least a couple of metres deep, with its shoreline about fifty metres from the start of our back lawn.

The whole flood event became an interesting revelation of the local geography. Those low ridge-lines, described at the start of this story, were shedding water down into the lower flatter land in between, the lowest points forming the many district swamps. With the amount of rain we had, those swamps quickly filled and overflowed, revealing that while there are no, or few indefinable creeks or drainage lines (or had been ignored and cleared and cultivated over) in the district, there is a general fall from southeast to northwest. This flood event was of such scale, that as each low area or swamp filled, the 'inter-dunal' areas became one big, interconnected flow of water, all heading northwest towards the Big Desert.

As our timber paddock was part of this, what happened next was quite dramatic, and for a while for Mary and I, somewhat alarming. It took almost two days to happen, but a big body of water was slowly moving from as far away as Lorquon, north-westwards towards us. When it arrived, the water level in 'Pine Flat' started to quickly rise.

As he had had a first-hand experience of the 1930 flood, Dad was first to call us to see if we were all right. As it happened, the township of Nhill itself was the centre of a flood emergency with several main street businesses being swamped by water, but his interest was our situation. When I advised him of our new small lake, but it was really a fair way from the house, he said "It was a lot deeper than that back in 1930. "It very nearly came up under the house!" "You better keep watching it." As it had always seemed to me that our house was on a bit of a hill, I remained sceptical, but we did get a wooden stake, painted it white, and put black tape around it to mark each inch of height, and hammered it into the back lawn – our flood gauge.

This might not sound an unusual situation, given that as I write this Australia has just been through a year or more of record flooding. But if you went out to Perenna now you wouldn't find a lake of water, and if you saw the location of that flood gauge, you would find it quite fanciful. How could such an apparently dry place, with no actual creek or river within thirty kilometres, ever have been subject to flooding?

But we were witnessing a reality. Mary and I sat on our back doorstep that night, with our prized new TV and most of our electrical appliances up on the table or benches, and gazed at the flood gauge, as the level visibly rose.

Figure 22 1974 Flood - the view from our back door

And it did get under the house, but well below the floorboards, before finally peaking, at which time it was overflowing across the paddocks to the north, with its final destination being a big limestone flat just south of the Netting Fence at the edge of the Big Desert. This became a quite large expanse of water, but it did recede more rapidly than other flooded areas such as ours at Pine Flat – our observation being that it was actually an aquifer recharge site. It was possible in the following days to go out there and see air bubbles coming to the surface as the level receded.

Thus, the house was safe and given our record crop was all safely in the hands of the Grain Elevators Board, the flooding was probably a net positive. It was really quite an exciting time to see all of the new water bodies throughout the district. We had our own private lake at our doorstop, and the build-up of water in the soil profile probably benefited crops for the next two or three years, or more.

Was there a negative? Yes, it was all that clover hay! For starters, given the blue tarpaulins were not in place to protect them, that seven and a half inches of rain in a few hours did a very good job of completely penetrating the stacks, thoroughly wetting most of the pressed hay and basically ruining it. It became sodden, mouldy and black, and even' fly blown'. Its likely use as fodder gone. But worse, Dad's warning that we may not have chosen the wisest place to stack half of it in the timber paddock proved correct!

With the water at peak level, the stack was in almost two metres of water, with the result that it slowly started to disintegrate and float away. No matter how ingenious the stack hand machine had been in alternating layers of bales in different directions to bind it together, the whole thing fell apart within a few days, drifting around in the trees, and washing up around the perimeter of the receding shoreline. Not only was it an economic loss, but also it made a mess of our otherwise beautiful new lake!

And one other negative. Given that the outfall from our septic tank system was now underwater, for our own health and safety, and our propensity to want to paddle across, and to swim and cavort around in the water, as the hot summer progressed, we had to revert to a makeshift hessian-clad long drop dunny, in the mallee trees out west of the house!

Figure 23 Myself with Bert our Golden Retriever, floating amongst the hay

The Lake was Full!

Figure 24 Outlet Creek flowing north from Lake Hindmarsh - early 1990's

La Nina didn't just dump this amazing summer rain on us in early 1974 and go away. As mentioned earlier, 1974's rainfall near enough to equalled 1973's record rainfall. As a result, we had another bumper year for crops, and in fact the benefit of all this moisture carried forward probably through the next two years as well.

But the other big consequence of all this rain was the amount of water carried down the Wimmera River, filling Lake Hindmarsh, which in turn caused Outlet Creek to run further north, filling Lake Albacutya, and then further on, to fill most of the 'terminal' Lakes in Wyperfeld National Park. This was truly a once-in-a-lifetime event, and I consider myself fortunate indeed to have witnessed this, and to be able to say that I have swum and snorkelled in beautiful almost crystal-clear metres deep water in a section of the creek just south of Lake Brambuck. A memory to look back on of a special time and place.

And while this was indeed an historically rare event, with most of the water drying away reasonably quickly, with Lake Albacutya dry again by the early 1980s, Lake Hindmarsh remained at good to sometimes overflow level right through until the late nineties, as apart from the single record dry year in 1982, rainfall remained generally around about average through that period of time.

The years immediately following the flood were good for us. La Nina abated, but we still had good seasons, with more 'normal' weather patterns. Crops and pasture grew well. We expanded our business significantly, leasing some neighbouring land, and then purchasing two more blocks. We upgraded our equipment progressively to allow us to work more land. The time of increased mechanisation with higher horsepower and larger capacity machines had arrived. In a decade, our main tractor grew from ninety horsepower to two hundred and fifty. We were not alone in this change, with the net effect being less labour was required at a district level. The district's population was in accelerating decline as a result.

Social life was changing but was still active and busy. As we saw it, the local district still had lots of young families, but the centre of activity became Nhill, as Netherby and Lorquon declined. Closure of the Perenna School in 1976, meant our kids went to Nhill for school. Increasingly our social life became much more centred on Nhill, rather than Netherby or Lorquon. The Netherby Football Club folded in 1983, while the Lorquon Cricket Club had earlier relocated its home ground to the Nhill High School Oval – somewhat influenced by having several of its players on the teaching staff. The other appeal of this move is that the high school oval was watered and was actually green though summer.

With the demise of local sporting clubs, a lot of the old social activities based around the clubs went with them. The Saturday night dances, and cabaret balls organised by the Footy Club were no more. For us, the gap was filled by Nhill based clubs and activities, particularly the Jaycees Club, which appealed to me as it was both a social outlet and more a self-improvement movement than a service club. This probably filled for me a part of the gap left by not going on to tertiary study. There was a strong emphasis on formal organisational procedure, and in developing skills in public speaking and debating. I found the latter to be stimulating and enjoyed the challenge of competing in a state and then national debating competition culminating in a national final in 1981, in Townsville, North Queensland. That was a good reason for an extended family holiday and road trip to the Sunshine State. Another upside to this group was the social side – we met interesting people whose background was not from a small rural district, and we talked about more than football and tractors! It led me down new pathways learning from people – such as all-night games of Chess.

And a key aspect of our life beyond the farm was, as mentioned, having Lake Hindmarsh, more or less permanently full, just a few kilometres away to the east. It was our summer playground, added to by the formation of a Sailing Club – largely a grouping of several families, Mary and I and the kids among them, who saw the lake as being really quite ideal for 'off the beach' sailing catamarans. These craft were well suited to the shallow water of the lake; although I must say, it was also highly suited for use of speed boats and water -skiers, who usually outnumbered the sailors, as well as many fishermen who just wanted to punt along in their tinnies. There was more than enough room for all, with an expanse of water over twenty kilometres long and ten kilometres wide. It really was a recreational playground for several months of the year. This didn't just apply to day visitors – with hundreds camping at Four Mile Beach near Jeparit, and also at "Schultz's, Beach" on the Western Shore for weeks on end through the summer.

Afternoons spent on the Lake often followed on to evenings there, sitting around a fire, cooking up a meal while watching the light fade across the water. Watching

Figure 25 Hindmarsh Sailing Squadron at Shultz's Beach

the full moon rise out of the sandhills beyond the Lake's eastern shore, shining across the water, is an abiding memory of many enjoyable times spent there.

I did enjoy the decade or so of the activity of the 'Hindmarsh Sailing Squadron', which filled both a social and a competitive sporting gap post the local cricket and football clubs. Given the expanse of water, it was easy to set a nice long triangular course, marked by buoys, around which we sailed regular races through the summer months. Of great assistance to us was the Nhill State Emergency Services (SES) group, who had a need for a regular training exercise for their rescue boat and crew, so this fitted in well that they managed our races for us, timing each boat around the course, enabling us to calculate both 'Line Honours' and handicap winners. They were also on hand to assist if required for the inevitable capsizes as we sought to maximise speed on windy days. It was a good arrangement, that only ended as the Lake dried up in the late nineties.

Figure 26 : I enjoyed sailing on Lake Hindmarsh

More Meetings

As the population declined, and local sporting clubs folded or retreated to Nhill, the local halls may have fallen into less social use, but they continued to serve an important purpose - as the location for the meetings that still took place as part of the conduct of the business of the many community organisations that remained. I have no doubt that one of the hallmarks of local rural communities is the commitment people make to the activity and success of local organisations. It was rare indeed to turn up at the Netherby or Lorquon Hall for a meeting and not find a whole lot of cars parked out front and around the hall, with heaps of people inside.

One organisation that met regularly was the Victorian Farmers Federation (VFF), Netherby Lorquon Branch. The VFF was, and still is, the peak farming industry association, existing mainly to act as a lobby group seeking to influence state and federal government policy. The way you could pick it was a VFF meeting would be by observing the twenty or thirty cars parked out along the road and noting that there would be one or two parking spaces left vacant right in front of the Hall. These were for either the local member of parliament, or more importantly, the local grower member of the Australian Wheat Board, who were generally expected to make an appearance. These gentlemen (always men, never women in my time!) were of sufficient importance to not have to walk too far to make their entrance as the meeting progressed. The AWB representative's Benz, or the MP's Statesman or Fairlane did warrant pride of place!

Long before the age of instant access to information that we have now, I well recall having my notebook and pen ready to note down the expected next 'pool' payment by the AWB. This would be provided by the AWB man, a Mr Jim Nuske:

"Gentlemen – I am pleased to be able to tell you that we expect to make a third advance of 15 cents per bushel before the end of September, and there might be as much as a further 10 cents a bushel left when we wind the pool up next March."

It was worth leaving a vacant car space to be able to go home armed with that information about our cash flow for the rest of the year.

The VFF was a valuable forum as I look back, and as the years passed. I did take the opportunity it afforded to participate in activities and conferences beyond the local level, and to assume roles of chairperson for both the Wheat Research Foundation of Victoria, and the advisory council of the Victorian Institute for Dryland Agriculture. These fitted well with my ongoing interest in scientific research and development, providing a window to some interesting people and research work.

Another meeting that most of the district turned out for (the men that is!) was the annual meeting of Netherby Fire Brigade. For many years, this was a two-pronged affair. Its late spring timing was carefully managed so that it would be a good afternoon to do the annual town burn off, prior to the meeting.

Long before such a thing as a rotary slasher that fitted behind a tractor, or a ride on mower, or even a 'Victa' mower was invented, it was a reasonable and necessary preparation for the oncoming summer to remove the drying grass and weeds surrounding our farmhouses and buildings by burning it off in a controlled manner. The other justification for this was "So that we could see the snakes" as Dad, and Jack Dodds used to assure us.

My first memories of this were that on a suitably still late afternoon, Jack would get an old hessian wheat bag, and hook a piece of wire to it, pour a tin of kerosene on it so it would light up easily, and then drag it around the outside of the garden fence, and over around the sheds, lighting up the dry grass, while Dad walked behind with a knapsack spray unit to stop the fire at its intended edge. We kids were instructed in how to bash the fire out with a well-chosen tree branch with lots of leaves on it, as a back-up. It was actually good fun, with the crackling and glow of the fire as the evening drew on, and the sunset was filtered through a blue haze of smoke in the mallee trees to the west.

This practice must have been a success over the years – we never had a serious fire close to the house or buildings, and we never got bitten by a snake. The downside was that the house was surrounded by nothing but bare, dry dirt and dust for the next six months, adding to how hot and exposed we felt as the summer progressed.

On the night of the fire brigade meeting this exercise was repeated, but at a somewhat larger scale. Everyone met at the Netherby fire shed, where a full complement manned the Austin Fire Truck. There were many hands left over, so they were all equipped with firelighters and/or knapsack sprayers, and a few others had their Utes with a forty-four-gallon drum of water and a fire pump in the back.

In no time at all there was fire everywhere throughout the township – every square yard of dry grass that wasn't in someone's house yard was targeted. Along the roadside edges, under fence lines, around the back of the old blacksmiths shop, down to the silo yard, behind the churches, and over towards the footy ground. It was always a well-controlled burn, with particular care taken to hose down wooden fence posts, and areas under the few scattered Black Box or Buloke trees left in the town where branches might have fallen.

By nightfall, all the open space in the town was blackened. It was safe to assume that there would be little by the way of fire hazard for the forthcoming summer. Like at our place, all that would be there would be bare, dry dirt. Time to repair to the Pub for a bit of refreshment, and then the Fire Brigade meeting at the Hall.

I well recall one such burn-off and following AGM – I was actually the secretary at the time.

'Chop' Nottle had just completed his first year as Brigade Captain, having been appointed at the same meeting a year ago. A key part of the meeting was the Captain's Report on the past year's activity. 'Chop' was able to report that it had been a good year, with only a few incidents that the brigade had to attend. There were no serious fires to report. The fire truck's maintenance was up to scratch, and it was always taken out for its regular weekly run to check it all out. The condition of the Big Desert fire break (pretty much along the Netting Fence) was a concern, but that was outside the volunteer brigade members' remit, being the responsibility of those further up the chain of hierarchy in the CFA.

'Chop' was pretty much into his report when a late entry into the Hall was made by his immediate predecessor, Jack Launer, who had handed over the captain's role last year after some twenty years in the job. As such, Jack's entrance commanded a bit of attention, and he took up the moment of quiet:

"Mr Chairman, I'm sorry to interrupt, but I'm late because there's a burning strainer post down at Bill Steven's corner. I've been trying to put it out with my knapsack, but it's still well alight. I think we had better get the Fire Truck out and go down there and put it out."

'Chop's' response was pretty quick: "Bugger the bloody strainer post! We're having a Fire Brigade meeting here! We'll go and put it out when we're finished!"

He went on to finish his report, and the chairman then declared all offices vacant, calling for nominations for the various positions. Due process was followed, with votes taken to confirm nominees as appointed to their positions for the following year.

It was only after the completion of formalities that the Fire Truck was taken down to put out the burning strainer post, or what was left of it. And I don't believe I was tasked as secretary to fill out an incident report to be sent to CFA headquarters for that particular 'turnout'.

Working the Fallow - Seriously!

Armed with bigger tractors, associated bigger, wider cultivation machinery and 'air-seeders', the years following the 1974 flood were our years of "peak" cultivation. Partially as a means of seeking to fully utilise the capacity of new bigger and more expensive machinery, but also in response to low wool prices, we transitioned our enterprise to a "continuous cropping" regime. This meant that every available acre was sown to crop every year, displacing the earlier system of clover-based pastures and long-term fallow. Grain legume crops and later oilseeds – Canola, were introduced as rotational crops with wheat and barley. We sold out of our breeding ewe sheep flock, using agisted grazing and the occasional foray into short term purchase and subsequent sale of sheep, to graze down and reduce crop stubbles through summer and early autumn. And we cultivated! Every time we received a shower of rain, it was out to "work up" more paddocks, and to re-work previously cultivated paddocks. Given the equipment and knowledge available to us at the time, most of the previous year's wheat and barley stubbles were burnt off prior to cultivation.

Perhaps here I need to mention some more reasons why we largely moved out of running lots of sheep. The main one was that our extended and spread-out enterprise did not lend itself easily to the movement and handling of sheep. It wasn't a farm set up with internal laneways and well-designed sheep yards. To get sheep around the farm, and to our shearing shed and yards meant droving them along the public roads, sometimes several kilometres at a time. Consequently, moving sheep meant you ran the gauntlet of having many cars, probably the school bus, a few trucks, and most likely a neighbour with his tractor and cultivator bar, driving one way or other through your mob. Hence, it was a regular happening to have hundreds of sheep split into several mobs along the road before you could get the dog to gather them back - if he could - or to have them push over a fence into an adjoining paddock.

And then there was shearing time. Mostly this went well, with local blokes (fellow cricketers and footballers) doing the shearing for us. But one year we had to rely on a couple of "professional" outside shearers. One was a really good bloke, but not his mate. He soon showed a doubtful temperament, mishandling sheep, who in turn would struggle even more, making his handling of them rougher still. It wasn't pleasant having that kind of worker, but shearers were in short supply, so we grinned and bore it.

But the clincher was his exemplifying the strongly sexist nature of the industry at the time. On the first day of shearing, Mary arrived at the shed with a basket full of lunch for morning 'smoko'. I recall she had Lisa with her as a toddler. She sat Lisa on a wool bale and proceeded to get the lunch out on a little table while the shearer's finished the morning run. Mary stayed a while to see if what she

Figure 27 Our trusty Sheepdog 'Blue' at work

had brought was all okay, before heading off to prepare for the next. The mood in the shed seemed tense.

As soon as she left, we got a tirade along the lines of: "I thought you blokes would know better! There's no bloody women in the shearing shed! You tell that wife of yours to just leave the lunch on the landing, and stay out of the shed from now on!"

This was the early seventies - I'm glad to say the industry now is almost completely opposite, with women making up a significant portion of the shearing and wool handling workforce.

Back to stubble burning, an interesting subject in itself. The earlier varieties of cereal crops that were grown through the fifties and sixties grew quite tall – in good seasons up to a metre or more. And the harvesting equipment of the time was designed to cut off and process the least amount of the bulk of the crop as possible – simplifying the threshing and grain separation process, allowing more area to be harvested in a given time. As a result, the subsequent stubble was often tall, thick, and dense. A heavy wheat stubble carried an enormous amount of fuel for any fire that might take place, be that accidental – luckily for us a rare event – or deliberately, as was the case in burning off.

Watching big wheat stubbles burn was a spectacular aspect of each autumn, both on our farm, and throughout the district. This was by necessity done with considerable care. Firstly, a really good fire break had to be established around the paddock edge, where those 'Hay Tracks' talked about early had another use.

These were scarified up via one or more passes with the tractor, leaving a bare earth boundary around the stubble. You needed to choose a good day with a light consistent breeze, and then hook a single section of a harrow up with a suitable length of wire to the back of the Ute. Dragged along the stubble edge beside the firebreak the harrow would fill with dry straw as you went. So, you picked the best corner of the paddock to start, with the wind blowing towards you, threw a match into the straw filled harrow, and drove slowly along, creating a wall of fire which burnt slowly upwind away from the firebreak. If you had your starting point correct, the main fire action wouldn't take place until you drove along the final edge – at which point the fire would rapidly grow with increasing heat and size as it burnt downwind towards the edges already alight, consuming everything as it went.

It was quite an adrenalin rush to be driving or sitting in the back of the Ute watching this all unfold. The noise and shear energy and power as the fire reached its peak was sometimes awesome - with a great pillar of smoke rising up above you. And it was bad luck for any animal such as a rabbit or hare (or mice) that might have thought the stubble was a nice place to shelter – although better luck for any local hawks or eagles who would quickly gather above, ready to swoop down for a feed.

It was equally bad luck for any remnant tree that might be standing in the middle of the paddock. Precious few of them were left on our farm, no doubt casualties of stubble fires past.

All that was left was a bed of black, smouldering ash. Depending on where the paddock was in relation to the farmyard, the next likely outcome, subject to the following days wind strength and direction, was a house full of fine ash and dust!

Such colossal stubble fires became less common as newer high yielding cereal cultivars were deliberately bred as 'dwarf' varieties – with much shorter straw. The plant breeding principle here was to encourage less vegetative growth, leaving more resource available for head development and grain production. This was a major development of the late seventies, contributing greatly to the much higher crop yields now achieved. A parallel development was the introduction of much larger and more powerful harvesting machinery, which was able to cut off and process much of the bulk of the crop. The stubble left standing behind was much less bulky, less likely to carry a huge all-consuming fire.

This meant even more time out on the tractor, using a big wide set of hydraulic harrows, dragging up and forming windrows of this much shorter stubble that could then be burnt to enable the next cultivating pass to proceed without straw blocking up the cultivator bar.

And that mighty set of harrows was then used again to fine down and level off the soil, so that a 'pre-emergent' herbicide could then be applied to the surface with a boom spray. This would prevent weed seeds from germinating, and provided that you placed your seed at sufficient depth below this layer when you then planted the crop, its first several weeks of growth would be free of the weed competition that had constrained yields in the past.

That being so, the whole farm was cultivated and or harrowed multiple times in the weeks up to seeding, often at speeds of 8-10 kmh. We spent months out "working the fallow" between harvest and cropping time. We didn't recognise it at the time, but as I have touched on earlier in this tale, all this rapid and repeated cultivation was leaving the soil very fine, with little 'structure', while the sub-surface was becoming compacted by the repeated passes of big, heavy equipment.

But all our effort was being rewarded by good crops, and good times, and the risk associated with this system of farming remained hidden from view – until the 1982 drought came along.

The 1982 Drought

Our run of good years came to a sudden end in 1982. The record shows that this was the driest year ever recorded at Perenna - just 145 mm. With little rain through the growing season (April – October), the only way a crop could survive and be worth harvesting would be if it was able to access sub soil moisture – that is moisture retained in the profile from the previous year. As it happened, 1981 was actually quite wet through winter, so there was moisture stored at depth. But hardly any of our crops managed to develop a root system able to benefit from this moisture. Why? The answer turned out to be another result of all that fallow working, and one of the many substantial lessons learnt from the 1982 drought.

The issue was that every time we cultivated the ground, we actually were drying that surface layer out. By preventing weed growth – the main objective of all our cultivating – we were preserving the moisture at depth, but we were drying out the top. This is fairly straight forward in hindsight – you can see the same effect if you fork over some soil in your garden on a sunny day – the inverted, broken-up soil dries out in the sun and wind. Therefore, you need to sprinkle your bed with the hose to allow any new seeds you plant to germinate. All of our paddocks were the same – by constantly drying out the top layer every time there was a scant shower of rain, we were preventing what moisture we may have had from being retained or connecting up with moisture below.

Consequently, the crops we did manage to get to germinate that year 'hung on' through the cooler months, but with very few 'sprinkles' of rain were not able to develop a root system that could penetrate down to the moisture in the soil below. Failure or very meagre grain yield was the result. Only a small percentage of the several thousands of acres we planted that year was worth the effort of harvesting. A key aim was to get enough"' grain for seed for next year's crop.

There were some important lessons to be learnt from the drought of '82. And this didn't just apply to us. The key issues were that leaving the soil surface bare and too finely structured left it significantly exposed to the risk of wind erosion - and indeed water erosion when the drought broke, and that we were wasting too much of our precious moisture by our constant cultivation.

Figure 28 Planting the crop - 1982

This record dry year did precipitate, pardon the pun, significant change to the way we, and many of our peers throughout the Wimmera Mallee (and beyond) farmed. Central to this was the widespread adoption of zero tillage and stubble retention. This form of farming is now almost universal in the region, greatly aided by technological developments such as precision seeding with minimal soil disturbance and GPS guidance. The ability of today's farmers to accurately drill this year's crop between the rows of stubble from last years is and was an amazing technological development.

Any sudden change you may make to any system of work or endeavour is bound to have unforeseen problems. Our practically immediate response to the 82 drought was to cease burning or removing stubble, which we thought we could achieve by the use of a 'chisel plough'. In theory this large, heavy 'wide clearance' machine had sufficient distance beneath its frame and between its tynes to allow stubble to pass through without blockage, which seemed to work when tested while conditions were dry. But once we had autumn rain to wet things up, the machine had major issues with blockage, leaving big lumps of straw and soil across the paddocks. Seed placement was poor, leading to less-than-ideal germination, and we also experienced what was referred to as "nitrogen lock up" - caused by the burial and sudden decomposition of crop residue making less nitrogen available to the new crop. As a result, our crop yields in 1984 were significantly depressed, and the following year was not much better. (We did have a bumper post-drought crop in 1983, when there was no crop residue to deal with)

We were not the only ones experimenting with new methods, with similar unintended consequences. One of the key region wide responses to the drought and its aftermath was the formation of the Wimmera Conservation Farming Association, which I was pleased to be part of. This was a forum which met regularly to discuss and exchange ideas and experiences in seeking to move to a more sustainable system of farming. This to me was a good sounding board – many others were making similar mistakes as we all found ourselves on a steep learning curve in seeking to radically change the way we farmed. Also influential was the availability of private agronomic advisors who were quite often from a background of government funded research and or extension roles. They were able to advise on a range of crop establishment and management techniques, having experience of what worked and what didn't across their clientele. We were well served in that space by some good advisors.

We steadily improved our minimum tillage equipment and methods, fitting the chisel plough with ultra-narrow 'points' which provided minimal disturbance of the soil and stubble, and good seed placement. This meant that we were no longer drying out the topsoil – our crops were more likely to germinate and be able to quickly develop healthy root systems that allowed the better utilisation of the available moisture. In the absence of cultivation, this new system of farming relied on chemical herbicides to control weeds, so a sizeable boom spray became a key element of the farming tool kit. The era of chemical use in a major way had arrived. We adapted our fertiliser application - lots more applied nitrogen - and crop rotation for this new system, and by the early 1990s were recording crop yields beyond any results we had in the very best years previously. All achieved while spending much less time out "working the fallow"! In fact, zero time working the fallow – it was a momentous change.

And it was a change driven by an increasing understanding of our fragile environment. It might seem a contradiction in terms, but the 1982 drought was for me a "watershed moment". We had to take a new approach to our environment – and we had many fellow travellers who shared this view, and new forums sprang up to assist in awareness raising and discussing what we could be doing differently or better. The Wimmera Conservation Farming Association which focussed on farming methods wasn't the only new group that formed. Other groups that I am proud to have been involved in establishing were the West Wimmera Farm Tree Group, the local Netherby Lorquon Landcare Group, and the wider Hindmarsh Landcare Network.

Figure 29 Scott, Lisa and Russell checking the Canola
(Photo by Brett Wheaton Photography)

A significant spin-off of the much-reduced time spent "out working the fallow" was that it did open up the opportunity to become more engaged in other pursuits beyond the day-to-day operation of the farm. For me, this was firstly an opportunity that the new world of computers opened up. We didn't refer to them as "IT" or Information Technology back then in the early nineties. I found myself spending some time with a local entrepreneur in Nhill, Kevin Tait, who had established Nhill's first computer outlet. Together, we developed a set of "macro" enabled spreadsheets to do basic farm budgeting and financial recording – a computerised multi column cash book – and found interest in this to the extent that we had over one hundred clients across the Wimmera using it. This was both stimulating and interesting, but given the rapid development of computers and associated new software packages and providers, this didn't develop into the ongoing business opportunity that Kevin, and perhaps myself, had imagined.

What did develop into something larger for me, and indeed eventually a new career beyond farming, was my interest in the natural environment, and how our farming experience had firmed up a conviction that sustainable use of the natural environment was a first order issue.

This really started one night at a meeting of the West Wimmera Tree Group in Nhill when a letter from an organisation I knew very little about was tabled as part of the club's inward correspondence – seeking expressions of interest from people involved in farming to become a Wimmera Facilitator (Part Time). That organisation was Greening Australia.

What was appealing to me about the role described was its focus on the application of recent scientific research work into the low-cost establishment of native vegetation – by direct seeding. This seemed to me to fit in well with my existing interest in research work for cropping, and experience as a cropping farmer. It was all about establishing plants from seed.

I thus submitted an application to this organisation and found myself appointed as Greening Australia's Wimmera Regional Facilitator. This led off into a new direction as time went on, eventually State Manager, but really is only important to this story in as much as it eventually became a new career path entirely, bringing to an end my fifty-four years at Perenna, that time being what this story is about.

North of the Netting Fence

Figure 30 One of the last remaining sections of the Netting Fence north of Perenna - Early 2000's

As mentioned earlier, there was a geographic feature about living at Perenna that couldn't be missed – the low line of sandhills that formed the southern boundary of the Big Desert – along the Netting Fence. You could gaze out to 'the scrub' from the front windows of the house. It certainly dominated the view to the north from the Perenna Hill and was clearly visible from the Perenna School Ground. We school students used to refer to the highest scrub covered point visible as "Hermans Hill' as we ran around outside at play time – because one of the district's characters a Mr Hermann Strauss lived in a mallee block just on the "inside" of the fence, near this feature.

For most of our youth, what lay behind the fence, out "in the scrub" remained unknown to us. For some reason we never went there. Dad (CT) used to say that "I've spent enough of my time out in the scrub – it's just sand, and it's no good for anything", and never showed any interest in exploring it with us or showing it to us. I guess this was a reaction to the fact that his early childhood was spent living on a then mostly scrub covered block "out there on the edge of the scrub" and given the then imperative to clear the mallee scrub from the land. He was brought up at a time when men "battled" the Australian "Bush" as they sought to push it, and all the "vermin" contained therein, back from the new European farming landscape they were creating. It was an enemy to be overcome. Fortunately, they didn't take that battle north of the Netting Fence – leaving this massive tract of native vegetation intact. But we never went out there.

Apart from Dad's expressed lack of interest, another reason why we didn't venture out there was we lacked the means – other than walking. Any tracks out into the scrub were mainly through deep sand – which might be okay for exploring on horseback, but we didn't have any such animals on the farm post war, and certainly no "Jeep" or similar all-terrain vehicle.

It wasn't until I was well into my late teens before I ever had an opportunity to venture out there. When I did, I found the sheer scale of this large area of bush to be both very impressive, and with a beauty of its own. There is over one hundred kilometres of uncleared bush to the north, and even more east to west.

Accordingly, other than providing a backdrop of low blue hills on the northern horizon, the Big Desert provided little excitement, other than when it caught fire. The first big scrub fire out in the desert I recall was in 1959, when a fire burnt out there for several weeks, apparently ending pretty much because it ran out of fuel – it nearly burnt out the whole vast area. I well recall the big pall of smoke rising for days on end, particularly as it burnt close to its southern edge near us. And the red glow at night was eerie. I recall Dad saying that he hadn't seen a fire like that since the 1930s. The cause of that fire, and pretty much every fire since, was lightning from a "dry" thunderstorm. The main response to that fire was to wait until it threatened to come out into open farmland, with local fire brigades mobilised to tackle it there. There have been many fires out there since, but never on the scale of the 1959 fire. This is largely because authorities are now far more focussed on managing fire out there, using controlled burns to reduce fuel loads. If a fire starts from lightning or an escaped controlled burn they will mobilise with aerial and on ground equipment to burn back from control lines. As to whether this new management is benefiting the long-term persistence of the diversity of native plants, animals and birds is, I believe, an open question, as very little of the desert is now unburnt for any significant period.

Figure 31 Murrayville is 130kms that way! (looking NW from Hermans Hill)

It really wasn't until Mary joined me out at Perenna that exploring 'the scrub' became one of the attractions of living on the edge of such a wilderness, and one of the things we did, with the aid of a four-wheel drive vehicle, or those of friends. We enjoyed driving out to 'Chinaman's Flat', which is about 15 kms in from the fence, and sometimes further on. But the place we visited most – and still do – was 'Hermans Hill'. This point is only a kilometre in from the fence, so was easily accessible by walking. The view from its 'summit' is now one of the real attractions of the northwest Wimmera, which I am proud Mary and I helped to establish along with members of the West Wimmera Tree Group. It has recently been upgraded to be readily accessible. It has expansive views to the north across the scrub receding to a distant horizon, across the cleared farmland to the south where the social scene of Perenna once thrived, and east across the expanse of Lake Hindmarsh. We saw this as a special place and enjoyed many a sunset from the top of the hill.

On one such occasion, while enjoying a drink as we watched the sun go down with friends Roger and Joan Gordon, Roger and I discussed what a good adventure it would be to walk through the scrub from south to north. There wasn't wide acclaim for the idea from our respective wives, Mary and Joan, and the subject was dropped. But it arose again sometime later, at the time of Roger's fortieth birthday, at their property at Broughton. This time the idea being refined to a proposal to walk from the desert's edge just north of their property, through the scrub for ninety kilometres to the Murrayville Pub on the other side.

A few weeks later there we were – standing at the end of a track on the netting fence that led into the Desert, weighed down by backpacks that we expected held all we would require for a three day walk north through the scrub. This was one of the best, although perhaps foolhardy, things I ever did, along with my good friend Roger. It was a true wilderness experience – the tract we walked out into is actually recognised as 'Wilderness' in the Victorian National Parks hierarchy. There are no roads or walking tracks or any signs of human use, in an area of one thousand four hundred square kilometres. The only way into and out of this area is to walk. Which is what we did.

Figure 32 Desert Walkers. Roger is the tall bloke!

The 'perhaps foolhardy' aspect of our adventure was that we both had no previous experience of an extended bushwalk through such a remote area. We did do a fair bit of research and sought advice from people with experience in walking the desert, in particular local naturalist Barry Clugston. This was before hand-held or any form of GPS units were available, and the best map we could source to help in navigating was a 'Floristic Vegetation' map. The Big Desert is basically a big flat region, with a jumble of low dunes, interspersed by flat areas, so a contour map is of little use. And there are no tracks or walking paths to be followed. That being so, the best way we had to locate where we were was to assess the native vegetation we were in, and reference that to our vegetation map. But not being expert naturalists, that was not very reliable, so we settled into our plan, which was fairly simple: We would use our hand-held compasses, and walk north for an hour at a time, estimating our progress to be on average four kilometres per hour. Thus, if we did that for eight hours, we could assume we had covered at least thirty kilometres. If we did it for ten, we might be close to forty.

Any way, it didn't really matter where we were, as long as we kept walking north, we would eventually come out of the desert into open farmland. We could then look to see if we could spot the Murrayville silo, and there was our destination!

The other part of our plan was to manage our water use well. Our advice had been that we would need two litres each per day. Based on that, we packed five litres, making the assumption we would probably do forty kms per day – so we would really only be out there for two and a half days. Water is heavy to carry in a pack – five litres equals five kilograms. That would do for our two and a half days! Weight of our backpacks was another key consideration, and our advice had been to keep it to no more than fifteen kilograms for a walk, through sandy ground, of that distance. By the time we added a small tent and sleeping bag plus food, we were over that target weight pretty quickly.

With respect to communication, we made the assumption it wasn't that important, as we were only going to be out there for three days! But as a backup, Roger took a handheld CB radio unit. Our plan for a day two check-in was with another good friend and local pilot Gary Aitken. Gary's licence prevented him from flying his small plane far out over the desert, as there was no opportunity for an emergency landing out there. So, he volunteered to go up and circle above Lake Hindmarsh at five thousand feet which should put him in line-of-sight with us and give us a call. Roger kept his CB turned on all the second day, until it went flat. But no call from Gary who kept his part of the bargain. But we hadn't allowed for the fact that the CB signal wouldn't reach the 50 kms required. Or maybe the earth's curvature was too great! Thus, the CB unit was one item of extra weight Roger could have done without!

We didn't stress too much about this, as we were focussed on our task, and adventure, at hand. At the end of day one, we were reasonably certain we would have covered over thirty kms, so were fairly pleased with that, as we set up camp on a small rise and watched the sun go down as we cooked dinner over a small fire. And listened to the sounds of the desert at night. But we did have a concern – we had both already consumed more than our daily limit of two litres of water. We put this down to the day having been unusually sunny and warm for early October, and it looked like getting cloudy – might even rain a bit. We should need to drink less tomorrow.

Tomorrow came, and we set off north once again. But we soon found ourselves discussing how it really was just as warm as yesterday, and we really did need to drink lots of water. Our small supply was an issue – we were going to be very thirsty before we got to Murrayville. We had to resort to plan B. That was to turn right and walk in an easterly direction until we came to the Murrayville Track, which couldn't be more than twenty kms away. Once there, we knew that we could obtain water if we followed the track northwards until we reached "Big Billy Bore". This is about thirty-five kms south of Murrayville. As long as we came out onto the track and recognised where we were (south, not north of Big Billy) we would be alright. It meant that our ambition to walk the whole ninety kms across trackless country would be thwarted, but common sense had to prevail!

Therefore, we walked east for a few hours, and suddenly, there it was! The Murrayville Track. We really didn't know we were near it until we were practically on it. This is the nature of walking through this sort of country – there are no landmarks or hills that serve as a reference point, so without the wonder of GPS, we walked until we realised, we were there. That was the end of our true wilderness adventure – the rest of our journey would be along the track. And we were south of Big Billy as far as we could tell. Therefore, we headed north again - with empty water bottles – and we were very relieved to arrive at Big Billy. Bore water never tasted so good!

The beer at the Murrayville Pub tasted even better after we walked the thirty odd kms to complete our trek the next day.

It may have been somewhat foolhardy to head into trackless country with little prior experience, but it was a great encounter in many ways. We both had a shared appreciation of nature, which we were certainly immersed in, but we also found that a trek is a valuable life lesson in what really matters. In the most direct sense, the only thing that mattered was our ability to keep on walking. There was no other assistance or help if we couldn't! And if we had a thousand dollars in our back pocket, it was for now at least, just a useless heap of paper. We shared a lot about what really mattered as we walked along. The sad postscript to this is that Roger was diagnosed with incurable cancer not long after this and passed away a relatively young man. He was probably my closest friend. I am glad I shared that adventure with him.

From Blacksmith to Workshop

It's all long gone now, but a key location in Netherby's heyday was the Blacksmiths. This was in a big, rambling, multi-gabled, bush-pole and corrugated iron building, just north of the Pub corner. I'm not exactly sure when it closed as a business, but I believe it was in the 1960s, most likely at or soon after the arrival of State Electricity Commission (SEC) mains power in 1966. The building remained as an increasingly derelict reminder of its busy past, until it was demolished, I think in the late 1980s. That was a pity – because if it was still there today it would probably be on the register of Victoria's heritage buildings.

For me, as a kid, it was a place of wonderment. This was largely due to the quite elaborate system of rotating shafts and pulleys mounted up in the rafters, all driven and connected by leather belts, with the biggest belt being that connected to a big slow revving internal combustion engine located in a 'lean to' outside the main work area. Beneath or fairly adjacent to all these rotating shafts, variously mounted on big timber posts or concrete blocks, was an array of machine tools, all of which added to the capability to make and or repair all manner of things associated with the conduct of local farming and other supporting businesses. The owner of this place was referred to as 'The Blacksmith', and he and his workers had undoubted trade skills that are now largely lost.

I found it quite mesmerising to watch them go about their work, which involved a big forge as a centrepiece of the building, with a fan blowing in air, driven by a belt connected to a pulley above, fanning the glowing charcoal. Metal could be heated to red hot and beyond in the forge. And there were all the other belt driven machines – lathes, grinders, drill stands, and a crosscut metal saw. All whirring into action by the pulling of a lever to engage its belt.

Because of its capability, and also due to the fact that few local landholders had their own well equipped 'Smithy', this was always a centre of activity. Not only was it a place to bring in items and machines for repair – but it was also as a gathering place for a yarn while your job was in hand. Particularly so in the colder months, with the warmth generated from the forge and the big old engine and its reservoir of cooling water- which always seemed close to boiling. It was a comfy spot to hang around in!

But change overtook this business, which I expect was at its peak while the prime motive power and means of getting around was horses and buggies, and farm equipment was uncomplicated, such as wagons and ploughs, all of which could either be made or repaired there.

The age of automation brought with it new type of businesses such as Motor Car and Farm Machinery dealerships – and their associated garages and workshops, staffed by a new class of skilled workers and mechanics to service and repair their modern products. These businesses were mostly located in Nhill, not Netherby. As a consequence, the decline of the blacksmith probably really set in after the war, but I am sure the final blow to its business model was the coming of SEC power in 1966.

Suddenly, we were able to purchase and operate a whole range of tools and equipment not previously available to us – such as electric arc welders, power drills, air compressors, angle grinders, bench grinders, power saws – the list goes on. And with them came the need to learn new skills to use these new tools and equipment, and to have a proper place where they could be housed and used - the farm workshop. The workshop soon became the centre point of a farm's operation, and the decline in need for the local blacksmith was almost complete.

There was always a machine to be serviced or fixed, or something built or made. So, I spent a lot of time in the workshop we set up at Perenna, and as the years went on so did our kids, who almost always turned up there to see what was going on after the school bus brought them home. For many years the school bus to Nhill started and ended its daily run in our farmyard. This was precipitated by the closure of the Perenna School in 1976.

Our workshop was, I believe, a good example of one of the benefits of raising a family on a farm. There is a real blurring between what is work, and what is recreation, or just plain fun. And a place for learning.

There was some serious work and repairs carried out there – like re-welding and reinforcing the rear beam axle of the Axial Flow Header. That was the day that our employee, Michael Atwell, called up on the CB radio to say "Aw Gee!" "I think we've got a problem here!" I drove down the paddock in the Ute to find Michael still sitting in the driver's seat, but lying with his back firmly on the backrest which was seemingly at a 45-degree angle as he looked up at the sky through the cab window. He looked like he was sitting in the space shuttle waiting for take-off! The header didn't have properly mounted rear wheels anymore; and its rear frame was buried in the mallee dirt. The axle assembly had given up after bumping and bouncing over a couple of thousand acres of paddocks left way too rough by our first attempt at direct drilling, and it had crumpled in two. I remember well that day's repair work in the workshop, after craning the axle and the rear wheels onto the back of the Ute and bringing it in. And then taking it back out, jacking the machine back up to level, and re-fitting it. Happy to report it never broke and fell off again.

There were many other examples of such workshop jobs. When I look back, I realise that very often what seemed to be disastrous failures and breakdowns, holding up the timely completion of cropping and or harvest, actually became interesting challenges and pieces of work in their own right.

But then there was time for a bit of fun and working on projects together with the kids. Like building a go-kart powered by a Victa lawn mower engine, which I am sure led to Russell and Scott's interest, and subsequent careers, in the Auto

Industry. This was later superseded by a dune buggy, built out of a disassembled Holden Kingswood, its body being replaced by a frame welded up out of old bore pipes, largely completed over a fortnight by Russell when he was meant to be on 'SwatVac'.

Figure 33 A re-engineered Kingswood, with Russell at the wheel

As time went by, things started to get made, done or happen in the workshop for which my presence, guidance or assistance was apparently deemed as unnecessary. Such as using more of that bore pipe, and a fair quantity of the acetylene gas otherwise used for cutting and welding, to launch potatoes, raided from the kitchen, way over the top of the shearing shed and out into the paddock beyond.

And then there was the very fast Greenfields Lawn Mower.

A bit of background about the Greenfield. This was a trusty machine, our first ever ride-on mower, purchased soon after such things became generally available. And it was one of the best and lowest cost machinery purchases we ever made – as it contributed to the lifestyle we enjoyed at Perenna. Firstly, it enabled us to neatly cut the grass and weeds around the house and sheds, making the need for the annual burn off less important, in fact redundant. And that meant we no longer were surrounded by bare hot dry dirt right through summer. And it enabled us to be a bit creative in mowing a larger open area - my first golf driving range - down into the timber paddock, and then planting new native pine trees along the edge, slowly connecting the garden area around the house into the larger native area of the timber paddock. To that we added a network of mown pathways around and through the timber for walking, riding or go-karting. All made possible by the Greenfield.

Another bit of background is that the Greenfield wasn't a particularly fast machine – flat out, it wasn't much more than a brisk walking speed – like all ride-on mowers, because they are built for cutting grass, not racing down a road.

Until the day, when for some reason Mary and I were away, and Russell and Scott found themselves in the workshop, wondering what they could tinker with. They spotted a couple of used steel sprockets, one big, and one small, and a section of roller chain, still in serviceable condition, from one of our old header fronts, and noted that the sprocket's centres, were by happy coincidence, the same diameter as the drive shafts on the Greenfield. Hence, without making the story any longer than it needs to be, it apparently only took an hour or so for the Greenfield's main drive to be converted from a couple of sprockets that provided a 1:1 drive ratio, to a 10:1 setup. This meant that its theoretical top speed went from five kilometres per hour to about fifty!

The first hint we got of what happened next was the following day when we were back and found Scott to be moving a bit uncomfortably. An eventual look under his tee shirt revealed a heap of bruising and skin loss, leading to an explanation of what had happened.

A precis of it is that when first tested, the eight horsepower Briggs and Stratton engine was not up to the task of moving the new highly geared ride-on mower off from a stationary start. Plan B was to tie a tow rope to the back of the Toyota Ute, with the lawn mower driver holding onto the rope for a towed off start, letting go when the machine could take over on its own. The Ute could then be used to obtain a speedometer reading of the pace achieved by the Greenfield. There were apparently several successful runs at a speed the Lawn Mower was never designed for, before its doubtful steering accuracy led to loss of control, with Scott in the driver's seat. It was undoubted great good fortune that he escaped with just a few scratches and bruises. Despite the admonishment, and the direction that the mower's gearing ratio be restored to normal forthwith, Scott and Russell didn't lose their interest in fast motorised transport, and, in fact, Scott went on to be one of few fully accredited Test Drivers for Ford in Australia.

The modification of the Greenfield wasn't the only unauthorised activity that took place in our absence from the farm. Like the day Mary and I went off for a bit of relaxation at the Baker Golf course, with our friend from Yanac, Richard Morphett.

The Baker Golf Club is another of the many local institutions that has now receded into history, located on a Crown Land reserve about 10kms north-west of Netherby. Baker is a little district similar to our Perenna, located on the edge of the Scrub, once with its own state school. The nine-hole Golf Course outlasted many other sporting clubs such as Footy, Cricket and Tennis, being actively maintained by a committed group right up until about 2020.

Anyway, Mary and I enjoyed a walk among the black box and mallee trees with Richard as we hit a few golf balls, before deciding it was home time – the kids would be home from school by now. The trip back home took us through Netherby, so we decided to call in to the Pub for a bit of refreshment on the way. I recall noting several cars and Utes parked across the road at the Fire Station, which was empty with the Fire Truck gone. We went in and ordered a couple of beers, and in retrospect I noted a quizzical look on the publican Reg's face as he poured them. I took a sip, and decided to ask an ice-breaking question:

"So, what are all the vehicles doing over at the Fire Shed?" "Is there a fire somewhere?"

There was a moment of silence before Reg replied:

"It's at your place".

Our Red Falcon car probably had its fastest ever trip home from Netherby after a couple of very quickly consumed beers.

The fire was out when we got there. All we had left to do was find out what had happened. It turned out to be a bit of rabbit control work that Russell had decided to undertake down the timber after they got home from school - that somehow got out of hand. The method chosen was to pour a tin of petrol down a rabbit burrow, and throw a match in. Where could he have ever gotten such an idea?

The resulting very rapidly established fire extended well out of the burrow, into the surrounding dry grass. Fortunately, it was not yet summer, with the grass still partially green, so the fire wasn't an inferno. But apparently the best efforts to use branches to smother it were to no avail, so Scott was detailed to race up to the house to alert their sensible elder sister Lisa, for whom blowing up rabbit holes wasn't an after-school priority. Lisa's presence of mind led to her calling the fire brigade, before dragging out a garden hose to wet down and defend the house.

The fire didn't burn much area, with the aid of all that area mown by the Greenfield, and never threatened the house before the brigade's arrival and quick work to douse it.

I'm not sure if the rabbit population was reduced to any degree.

Crop Dusters

It wasn't until well into the 1960s that the use of chemicals for weed and pest control in our crops became a significant part of farming. Prior to this, as mentioned earlier, our farming methods were very close to what today is referred to as 'Organic'. The only artificial input was phosphate-based fertiliser, with some added trace elements. This system of farming accepted that there would be some weed growth in the crop. As a general rule of thumb, the less weedy a crop was, the better its yield, and vice versa. Thus, a weedy crop might yield 5-8 bags to the acre, while a less weedy or 'clean' crop might yield 10 or 12 bags to the acre, or even more. Apart from depressing yield, weed growth within a crop had another side effect – and that was the contamination of the grain at harvest time with weed seeds. Understandably, the eventual users of our grain didn't really want a heap of weed seeds as well as the wheat or barley they were buying and set an arbitrary limit on how much of a given sample of your grain could be 'foreign matter' or weed seeds, generally set at 5%.

This issue was mostly dealt with by what I understand to have been a peculiarly Australian adaptation fitted to the grain harvesters we used back then - most of which were designed and built in Australia. A "small seeds" rotating screen was fitted, through which the harvested grain would pass before it went into the machine's grain tank. As the weed seeds were generally smaller than the wheat or barley grains, most would fall through this screen and be diverted to a smaller screenings bin on the machine. For most part, this resulted in the grain sample being of acceptable standard to be delivered to the local silo.

But times were changing, and chemical herbicides that could enable weeds to be controlled within a crop were being developed and trialled – the first of which I recall being the "24D" broadleaf herbicide.

When this weedkiller first became available, the means to apply it fell to a growing band of fearless airmen whose machines, skill, and nerves equipped them with an ability to fly at low level up and back across a paddock at a hundred miles an hour or more, dispensing chemical as they went. We, and most other district farmers at the time were not equipped with such a thing as a boom spray, so we lined up as new clients of these "Crop Dusters".

Given that they were applying a liquid spray to our crops, the title as 'Crop Duster' seemed a bit odd. I expect it came from the use of planes to apply fertiliser, mainly in more hilly grazing country to boost pasture growth, prior to the new herbicide developments.

And so began a revolutionary change to the way we farmed – the era of chemical use had arrived. The benefit of having a crop duster come in and apply 24D to a broadleaf infested wheat crop soon became obvious – both in a cleaner 'sample' and increased yield at harvest time.

Most of the "Crop Dusters" came from other parts of the region or state, with the Nhill Aerodrome as a convenient and well-located base while in the district. I well recall an element of excitement, and in retrospect, a big element of insufficient personal protective care, the morning we first had a Crop Duster spray a crop. Our local agent having inspected the paddock and recommended the chemical to be applied, had provided the pilot with a rough map of the paddock and its location, and I stood out in the paddock, eighteen steps in from the fence as directed. After a time of peace and quiet, listening to the birds enjoying their morning, I heard the drone of a plane engine coming up from the direction of Nhill. So, I followed my instructions: "When you see him coming, start waving your hat so he can see you, and he'll know he's at the right place".

I waved my hat as the plane approached: he roared overhead and circled the paddock. Good! He saw me and dipped his wings before heading away to turn and line up with me for the first pass. I had been told "As soon as he's lined up for his first run, start walking another eighteen steps so you can wave him in for next run back".

Being very keen to know he was properly lined up, I made sure he was on-line as he came down and turned his spray on. I soon realised that a plane doing over a hundred miles an hour got from the edge of the paddock to its middle rather quickly. I reckon I was about one and a half steps into my eighteen when he roared straight over my head, missing me by about a metre! What didn't miss me was the cloud of 24D herbicide gently descending onto the crop. I learnt that it was best to start walking the next eighteen steps much more quickly! And thenceforth whenever I "marked" for a crop duster I wore overalls and kept my hat on, having something else to wave at the pilot.

The thought just occurred to me – is that why I had rather rapid hair loss from my early forties?

The shortcomings of relying on aerial spraying soon became apparent – there were only so many crop dusters around, so timeliness became an issue, as was accuracy, or rather lack of it. Accuracy wasn't necessarily important for the first type of chemical herbicides that became available – 24D amine and ester formulations were rather volatile – as long as you sprayed it somewhere near the weeds, they were likely to die! But off target damage was a definite risk, in addition to which a wide range of newer herbicides did need to be applied more accurately to be effective, and to avoid drift. This required application by hydraulic nozzles, ideally set no more the 50cm. above the crop - in other words a Boom Spray. Consequently, a new type of farm implement was added to the line up in the shed – and over time became the most used machine on the farm, as we moved to zero tillage.

But there was still a role for Crop Dusters – mainly later in the season when crop growth meant that driving across and through crops with the sprayers then available were likely to cause physical damage to the crop. And the introduction of crops such as grain legumes and canola brought with them new pests and diseases to be controlled later in the season – this was best done by plane.

For Mary and I, there was a spin-off benefit of this. When our long term nearest neighbours, John and Janice Evans and their family left in 1986, their farmhouse, about a kilometre away, became vacant, and with it a real feeling of change - we were becoming a bit lonely out there. I know it played on my mind, watching the sun set across the empty paddocks and the empty house and farmyard to the West.

Until along came a barnstorming Crop Duster pilot Fred Vanderol and his partner Arlie. Being able to land his plane and taxi it right up to just in front of the house suited Fred just fine, so it became his base to build his local clientele – including us. Our loneliness and lack of near neighbours changed for the better! Fred and Arlie were a very interesting couple, and we enjoyed their company. And he reintroduced me to chess!

Fred's aeronautical career started with the RAAF – he was quite nonchalant in his description of what it was like flying Sabre Jet fighters out of the Pearce RAAF base near Perth. One such story was quite amusing, recounting to me how he was admonished by his superiors for completing a "manoeuvre" too low and too close to the mouth of the Peel River, alarming local surf fishermen as he roared past at 700 kmh less than ten metres above the waves.

This probably was good training for his move into low level crop dusting, which he then turned to. Crop dusting is no doubt an adrenalin producing way to spend your working day. I got a bit of a look into this crazy, exciting line of work one day when Fred asked if I would like to see what it was like. My answer was yes, and five minutes later I was sitting on a metal cross bar, with no seat belt, in the cockpit beside Fred as we took off. We did a couple of dummy runs - no spraying - across one of our paddocks. I can still picture the rush of green as we zoomed along just above the crop, and then the centrifugal force either pushing me across hard against the cockpit wall, or tight up against Fred as we turned at each end of the run. Fred was at pains to point out that the turns were very much wider than he would normally do!

And, of course, with excitement came risk. Fred had already had a serious accident when he hit a powerline in Tasmania. That was how he came to meet Arlie, who was a theatre nurse for the plastic surgeon who treated the substantial burns he received to his face and arms.

Fred and Arlie didn't spend long as our neighbours, before the Evans Farmhouse became vacant once again – they moved to the Darling Downs region in Qld, where Fred saw greater business opportunity. We Wimmera farmers were investing in our own ever improving boom sprays, so Fred and his fellow ag pilots looked elsewhere.

Figure 34 Fred Jumpstarting his Crop Duster out on the front road after a cup of coffee with Mary

Before he left, he did quite a bit of work for us, including the day he sprayed our paddock just east across the road from his airstrip and their house. I was out in the paddock marking for him, as he did east – west runs. As we got to about the middle of the paddock, his turns were just about directly over their house. I took up position to mark his next run, but he didn't come zooming back. Instead, I stood and watched as he did several really tight turns around and around their house, his wing nearly pointing to the ground, before he came flying back, resuming the job, as I stepped out the runs.

The paddock was soon finished, so I had to walk back about half a mile to retrieve the Toyota Ute and drive home. As I turned the corner, I saw where Fred had parked the plane – on the limestone road, just in front of our house. He was inside having a cup of coffee with Mary. I joined them and had to ask why he had spent so much time flying around his house, and not spraying our crop. His answer was "I noticed that Arlie was out sunbaking on the front lawn – and she was topless!" I had to forgive him for interrupting the spray work.

After they left and moved to Queensland, we never saw Fred and Arlie again, and did hear that Fred had another accident. This time it was fatal. This was sad news – I'm glad of the time we spent with them.

Perhaps the final part of my story of life at Perenna should be about spraying, and the heavy reliance we came to have on the use of chemicals. A lot of the latter part of my time farming involved working with herbicides and pesticides.

Maybe my explanation of this is one of contradiction. On the one hand, I have no doubt that our crop yields and therefore profitability was significantly enhanced by our use of these, particularly as we moved away from constant tillage. I believe that not only were we a more profitable business, but we were also protecting our valuable land as well.

But I also recall standing in amongst a stockpile of used chemical drums, with a Ute-load of newly acquired ones, soon to be decanted into a sprayer, thinking "Is this all that farming has come to be? Am I going to spend the rest of my life emptying tins of poison?" Maybe this moment was influenced by our then use of a particular type of chemical – a nematicide applied to grain prior to seeding to control cereal cyst nematode (CCN). CCN was a problem that could cause very poor crop health and growth at the time, and nematicides were a last resort defence. This stuff was nasty and poisonous; thus care was really needed. But its nastiness seemed to also extend to the internals of the pump used to spray it onto the grain as it was being loaded, which failed with regularity. Having to dismantle an inoperable pump laced with this stuff was bad enough, but I think my real tipping point about its use was one morning walking over to the shed and seeing several dead sparrows on the ground. Now sparrows are a pest species, their loss wasn't really an issue, but what had happened to them was. Despite the greatest care to avoid grain spillage the night before, there was a small amount of treated grain laying on the truck tray. The sparrows could only have found this grain during the early morning and were already dead close by. If this could happen so quickly to this pest species, what about others that were not pests?

Fortunately, a combination of plant breeding and better crop rotation, and paradoxically, strategic use of herbicides, meant that CCN became much less of a problem, and nematicides were no longer represented in that heap of empty chemical drums. But the memory of not wanting to use it or spending my days decanting potentially dangerous chemicals lingered.

Looking back and reflecting that now well into my seventies I do not appear to have suffered any ill health that could be attributed to the use of chemicals on the farm, I believe that we were always as careful as we could be with the use of chemicals. And their use was, on balance, highly beneficial, and this continues to be so for farmers today. But not wanting to be always working with them was one reason why it was time in my mind to move on.

Conclusion

But the main reason why it was time to move on was really the core part of this story of life at Perenna. It just became too lonely out there. As already explained, my lifetime there spanned a time that saw a rural community decline and virtually disappear. Hardly anyone lived there anymore. This really became the key issue for us as the 1990s progressed. Lisa, Russell and Scott left for their education and careers - a turning point of course for any family. My increasing work with Greening Australia saw me away for days at a time. This left Mary to look after things around the farm on her own, which I know made her feel isolated. And I recall driving back home thinking "Gee, it's a long way out here". That would have been alright if we still had neighbours, or people in the area. But we didn't. Additionally, my developing work with Greening Australia provided almost a mirror opposite insofar as it involved interacting with lots of different people. Not just other landowners, but many others from a range of backgrounds and perspectives - city and country. I was finding it both interesting and stimulating – and a good cause. It was also a factor that I needed to be closer to the action if I was to further expand into this work.

We are herd animals – we are best when we are in proximity with others. My feelings here also might seem to be contradictory. One of the lasting memories of life at Perenna is of wide-open space, of a big starry sky without any hint of 'light pollution'. Of peace and quiet. In fact, complete silence in terms of human activity. That's how it was out at Perenna in the late nineties. But it was too complete.

At the start of this story, I explained that I don't know why my forebear Thomas Dodds left his place of birth with his family and travelled to the other side of the world in the pursuit of something different. And that I didn't really know what their day-to-day life was like.

I hope now that you do know why Mary and I and our family left my place of birth. I note that unlike Thomas Dodds, I only moved 100 kms. He moved 24,000! And I hope that you also know at least a bit about the community in which I lived for 54 years, and what our day-to-day life was like.

THE END

Appendices

Rainfall Chart, 'Perenna Lodge'

This is compiled from the annual rainfall records as measured daily at Perenna from 1930 – the year the house was built, until the year 2000 when Mary and I left, and the house became vacant, with no one there to take the daily or monthly recordings.

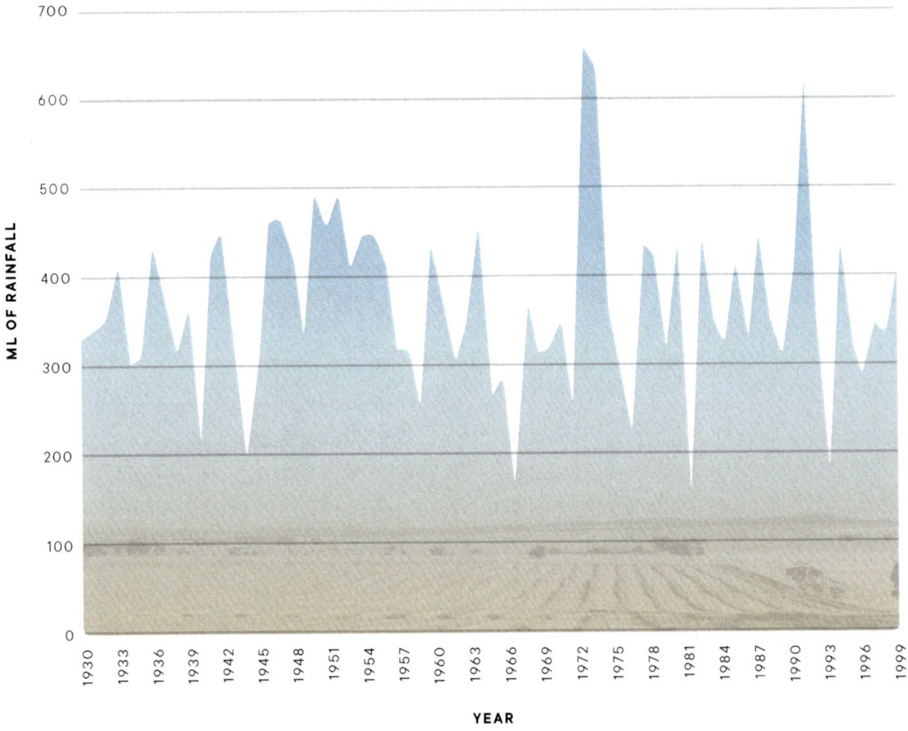

RAINFALL RECORD CHART - PERENNA LODGE

	JAN	FEB	MAR	APR	MAY	JUN	JUL	AUG	SEP	OCT	NOV	DEC	TOTAL
1930	8.8	2.8	17.3	47.5	53.3	81.3	52.8	22.3	33.0	5.8	5.3	3.5	**333.3**
1931	10.3	0.0	14.5	63.3	49.3	79.5	28.8	32.5	34.8	2.3	22.5	6.0	**343.5**
1932	0.0	55.0	33.5	45.5	23.0	42.0	37.0	35.8	19.5	28.3	10.0	24.0	**353.5**
1933	39.3	0.0	17.0	20.0	59.5	22.0	23.5	51.0	33.0	12.3	77.0	60.8	**415.3**
1934	8.0	10.3	6.8	39.0	0.8	6.8	28.8	32.3	38.3	79.5	44.5	9.0	**303.8**
1935	2.8	5.3	16.3	13.3	24.5	34.8	37.0	53.0	57.0	23.5	14.3	24.3	**305.8**
1936	58.0	1.0	21.5	3.0	46.0	53.8	104.3	35.3	8.3	37.8	3.0	65.8	**437.5**
1937	45.8	18.8	42.3	6.0	18.8	23.8	20.5	47.5	29.5	32.0	18.5	77.3	**380.5**
1938	112.0	11.5	7.5	51.0	0.8	37.0	51.3	10.0	9.0	3.8	10.0	1.3	**305.0**
1939	22.0	34.8	3.5	47.5	50.8	40.8	21.3	64.0	19.5	19.8	40.0	5.5	**369.3**
1940	12.3	3.3	6.5	68.3	4.3	5.5	27.3	13.5	26.5	7.3	20.8	18.5	**213.8**
1941	95.3	4.5	46.5	31.3	4.0	39.5	58.5	20.8	62.8	33.3	16.3	15.5	**428.0**
1942	19.3	17.0	3.8	30.5	66.8	79.5	31.3	71.0	46.0	43.0	34.5	13.0	**455.5**
1943	13.5	27.0	10.5	27.3	11.8	44.0	21.8	43.3	56.3	18.8	19.8	7.0	**300.8**
1944	14.0	4.0	7.0	19.3	40.0	1.8	26.3	5.5	8.3	27.3	11.8	22.0	**187.0**
1945	5.5	45.8	4.3	6.8	23.3	43.3	26.0	46.0	19.5	25.5	32.8	16.3	**294.8**
1946	56.5	94.0	51.3	6.8	23.5	43.3	59.0	30.3	18.3	21.5	28.5	32.0	**464.8**
1947	3.5	52.5	44.3	13.0	12.8	32.5	56.0	35.3	63.0	51.8	68.8	37.3	**470.5**
1948	0.0	10.0	1.5	66.0	30.8	52.8	37.5	21.0	20.8	71.0	52.5	59.5	**423.3**
1949	11.3	57.5	12.8	1.8	28.0	19.3	33.8	8.8	39.8	76.5	45.5	1.5	**336.3**
1950	76.5	103.0	20.5	96.0	27.3	22.3	29.5	51.3	27.8	32.8	13.8	0.0	**500.5**
1951	26.0	31.0	12.5	20.0	57.0	76.3	58.8	57.0	16.3	61.3	5.5	33.5	**455.0**
1952	65.5	25.3	11.3	27.0	85.5	60.3	30.0	45.3	42.3	34.3	60.5	12.0	**499.0**
1953	34.3	1.3	0.8	24.3	13.5	68.5	54.5	66.3	33.8	23.8	50.8	39.0	**410.5**
1954	48.0	0.0	4.5	115.5	13.3	15.5	34.8	61.0	9.8	51.8	24.8	67.5	**446.3**
1955	1.3	40.5	17.3	10.3	59.3	106.3	22.5	76.5	48.0	29.0	26.3	15.0	**452.0**
1956	15.5	0.8	52.8	34.5	74.0	42.8	59.0	29.8	41.5	52.5	10.3	7.3	**420.5**
1957	1.5	106.5	14.8	10.3	18.3	29.5	32.8	29.0	7.3	33.8	28.8	5.8	**318.0**
1958	5.8	3.0	9.3	5.3	50.8	5.3	58.3	57.3	27.8	64.8	29.8	3.5	**320.5**
1959	37.8	25.8	16.0	2.0	3.5	8.5	7.5	46.0	23.3	28.0	23.5	29.0	**250.8**
1960	39.5	32.5	17.8	44.8	63.3	26.5	31.3	35.5	61.0	5.8	64.3	14.5	**436.5**
1961	0.0	12.0	41.0	71.3	9.3	18.5	49.0	21.5	32.5	19.8	20.0	73.0	**367.8**
1962	19.3	8.0	8.8	1.3	40.8	42.8	13.5	32.8	49.3	33.8	15.5	38.3	**303.8**
1963	46.8	0.0	15.8	24.0	62.8	52.8	34.5	41.0	53.5	18.3	0.0	3.5	**352.8**
1964	5.3	2.3	5.0	45.3	17.3	35.3	50.8	39.3	76.0	90.3	72.5	16.3	**455.3**
1965	0.8	2.3	6.5	19.0	38.3	13.3	58.5	54.3	14.5	5.8	22.3	32.8	**268.0**
1966	19.0	15.0	7.3	7.3	8.3	33.8	45.3	20.3	37.8	33.5	11.5	50.5	**289.3**
1967	3.3	24.3	3.0	2.0	17.5	6.8	33.3	44.8	5.5	1.3	2.0	12.0	**155.5**
1968	33.3	6.5	27.3	34.0	56.5	31.8	38.0	48.3	13.5	46.5	22.5	16.5	**374.5**
1969	3.3	36.8	26.8	23.3	35.8	7.5	56.0	45.0	21.8	11.3	18.3	26.3	**311.8**
1970	19.0	2.0	19.3	88.0	15.3	13.5	10.8	37.5	19.8	9.8	63.0	17.5	**315.3**
1971	2.5	20.0	4.0	82.3	31.8	31.8	18.3	24.0	45.5	34.3	40.8	17.5	**352.5**
1972	40.0	36.8	0.0	30.3	7.3	12.5	25.0	48.0	13.0	10.3	24.8	1.8	**249.5**
1973	57.8	139.8	13.3	23.3	37.8	46.0	57.0	59.8	38.8	152.3	24.0	13.8	**663.3**
1974	195.8	20.0	36.8	56.5	78.3	19.3	43.0	42.3	48.8	97.3	1.3	2.5	**641.5**
1975	10.0	0.8	19.0	7.8	44.5	10.0	41.5	25.8	62.3	127.8	14.8	4.0	**368.0**
1976	18.0	34.8	0.0	17.0	4.0	27.8	13.0	17.0	32.5	39.5	64.0	33.0	**300.5**
1977	22.0	13.0	19.5	5.5	44.0	24.8	13.0	9.3	15.3	17.8	27.8	13.0	**224.8**
1978	25.8	3.8	21.5	43.3	29.3	40.0	61.8	51.0	74.0	27.5	50.3	9.8	**437.8**
1979	56.5	17.0	5.8	24.5	49.5	25.3	21.3	56.5	93.8	30.3	45.8	2.3	**428.3**
1980	2.0	0.0	6.5	40.0	34.0	42.5	47.0	33.3	21.5	69.0	7.5	11.5	**314.8**
1981	8.0	31.3	0.0	11.3	58.8	69.8	72.8	100.0	19.5	41.3	27.5	2.0	**442.0**
1982	18.3	6.3	4.3	16.3	11.0	25.0	8.5	8.3	11.3	5.3	11.5	19.5	**145.3**
1983	3.0	0.0	88.5	27.3	30.5	16.0	40.3	52.0	55.8	24.5	66.8	43.8	**448.3**
1984	25.5	1.3	61.3	7.5	6.8	9.3	50.0	101.5	33.3	6.3	39.8	8.5	**350.8**
1985	7.5	0.0	13.5	13.5	33.0	18.0	13.0	87.3	32.8	49.8	31.0	22.5	**321.8**
1986	1.3	0.0	0.0	37.5	42.5	12.0	71.5	74.3	29.0	71.3	13.8	63.0	**416.0**
1987	0.8	0.0	0.0	11.8	69.3	56.8	33.0	25.8	4.3	49.5	9.3	61.3	**321.5**
1988	28.0	19.0	66.5	11.5	87.3	63.3	37.5	16.8	40.8	14.5	22.8	43.0	**450.8**
1989	0.0	34.0	27.0	19.5	41.8	42.5	52.3	42.0	42.8	37.5	8.5	6.3	**354.0**
1990	0.5	40.0	16.0	6.3	5.0	41.8	78.0	62.3	22.8	13.5	12.5	9.5	**308.0**
1991	41.3	0.0	6.0	19.8	4.0	138.5	39.5	74.3	58.8	3.0	25.3	3.3	**413.5**
1992	31.5	11.0	19.5	46.0	84.3	31.8	30.8	84.5	77.0	45.5	45.0	118.3	**625.0**
1993	38.8	4.5	8.8	0.0	23.0	29.3	49.8	38.0	33.5	56.3	27.5	16.3	**325.5**
1994	18.8	7.0	0.0	0.0	25.0	43.5	25.3	8.0	12.0	30.5	11.9	0.0	**181.9**
1995	20.5	25.7	0.0	58.7	56.2	92.0	62.0	19.0	22.0	81.0	3.0	2.0	**442.1**
1996	13.0	21.0	6.0	18.0	22.1	71.9	60.0	38.5	65.0	2.5	3.0	3.5	**324.5**
1997	22.5	9.0	0.0	1.0	39.0	11.0	14.0	43.5	70.0	28.0	41.0	8.0	**287.0**
1998	6.5	35.5	4.0	69.5	15.0	30.0	39.0	40.5	46.0	26.5	22.0	15.0	**349.5**
1999	15.0	46.0	0.0	2.0	21.5	22.0	32.0	48.0	30.0	35.5	58.0	21.0	**331.0**
2000	4.0	100.0	2.0	34.5	13.5	23.0	64.0	30.0	49.0	30.0	61.0	0.0	**411.0**
AVG:	**25.0**	**22.7**	**16.3**	**28.9**	**33.6**	**37.0**	**39.5**	**42.0**	**35.3**	**36.1**	**28.2**	**22.4**	**367.0**

Mary's Genealogical Research

With the family bible as a catalyst or starting point, Mary has channelled her interest in genealogy, tracing our (Dodds) forebears back somewhat earlier than the information in this family heirloom. This is a precis of her work, going back six generations to 1796. She has an extensive trove of hard copy documents and photos from which this is drawn.

Figure 35 The Family Bible

First Generation: JEREMIAH DODDS - JANE EVENS

JEREMIAH DODDS

Born: 1796, Newburn, West Denton, Northumberland, England – parents William and Ann nee Evens. Ann may have died in childbirth the year after Jeremiah was born, (not confirmed)

Died: January 1857

Married: Jane Evens

The 1841 census for the United Kingdom shows that Jeremiah and Jane were counted at this address along with other members of the household listed below, and their age at the time. Jeremiah was listed as an agricultural labourer.

Nicholas, 23 years	**Sarah,** 20 years
John, 17 years	**Jane,** 15 years
Jeremiah, 11 years	**Thomas,** 8 years

Newburn is a village and a parish in Castle Ward district, Northumberland. The village stands on the River Tyne, near the Roman Wall, one mile north east of Ryton railway station.

In the 1851 English Census, Jeremiah and Thomas and parents seem to have lived at the same address as the 1841 census. Jeremiah Jnr was listed as a joiner and Thomas 19 years as a blacksmith.

Jeremiah snr. died in January 1857 at the age of 61, this being the year that Thomas his son, left for Australia.

Jane Dodds passed away on 12th January 1864 at the age of 68.

Second Generation: THOMAS DODDS - MARGARET FAWCETT

THOMAS DODDS

Born: 1833, County Northumberland.

Married: 1855, St. Johns Church, Newcastle Upon Tyne, England.

Died: 2 August 1885, Shorts Road, Coburg, Melbourne Victoria, Australia.

Buried: 4 August 1885, Melbourne General Cemetery.

MARGARET FAWCETT

Born: 1833 - New Castle Upon Tyne.
Daughter of Abraham Fawcett and Margaret Robson.

Married: Thomas Dodds.

Died: 11 July 1907, South Yarra, 25 Little Park Street.

Buried: 13 July 1907 Melbourne Cemetery.

Thomas, Margaret and Jeremiah are buried in the same grave which has no headstone.

CHILDREN OF THOMAS AND MARGARET

Thomas, Born 11 December 1855, Benwell, Newcastle Upon Tyne.

Jeremiah, Born 1857, Newcastle Upon Tyne, England, christened 6 June 1857.

John, Born 1858, Melbourne, Victoria, Australia – Died 1861, Melbourne.

Margaret Jane, Born 1863, died 25 June 1865 – buried Melbourne Cemetery.

Mary, Born 1866, Melbourne, married Albert King 1890 in Melbourne.

Thomas and Margaret's son Thomas was born 11th December 1855 in Benwell, Newcastle Upon Tyne. The following year their second son Jeremiah was born in 1857. Thomas and Margaret and their infant sons' left England in the Summer of 1857 on the "Ellen Stuart" and several months later arrived in Melbourne. Thomas' stated occupation being blacksmith.

John their third son was born in about 1858 in Melbourne and died at the age of three years in 1861.

Their first-born daughter Margaret Jane was born in 1863 but died aged two years on 24 June 1865 of scarlet fever and whooping cough. She is also buried in the Melbourne cemetery. Thomas and Margaret's address at that time was 4 Palmer Street, Fitzroy.

Mary, their second daughter was born in 1866.

Thomas (Snr) died in 1885 at the age of 52 years, of pneumonia. His occupation was still listed as blacksmith. At the time of his death they lived in Shorts Road, Coburg. He is buried in the Melbourne General cemetery (CC876) no headstone.

Mary married Albert King in 1890 and her address in 1891 was 129 Station Street, Carlton.

In 1891 Jeremiah Dodds died at the age of 34 years, his address being - Fitzroy. Jeremiah was married to Effie Oaks and his two daughters Maud and Alice predeceased him.

Margaret Dodds died on 11 July 1907 (aged 74) of the effects of gastric cancer. The place of death 25 Little Park Street, South Yarra - this may have been a hospice.

At the time of her death Thomas and Mary were the only surviving children.

Third Generation: THOMAS DODDS – MARY JANE RAINE

THOMAS DODDS

Born: 11 December 1855, Benwell, Newcastle Upon Tyne.

Died: 23 March 1932, Alfred Hospital, Melbourne.

Buried: Fawkner cemetery.

MARY JANE RAINE:

Born: 13 September 1860, Collingwood, Melbourne.

Married: 19 March 1877, United Methodist Church, Hoddle Street, Collingwood.

Died: 12 August 1942, Corpus Christi Hospital, Kew.

Buried: Fawkner Cemetery, Melbourne.

CHILDREN OF THOMAS AND MARY JANE

John Thomas, born 7 July 1879, 75 Oxford St, Collingwood, died 3 March 1880.

William born 23 April 1882, 3 Collingwood Terrace, married
Elsie M Hughes 17.9.1902, St Kilda. Died 4 September 1934, 28 Plant St, Northcote.

James Ridley, born 4 July 1883, Gipps St. Collingwood, married Mary Jane Sutherland 2 January 1903, Collingwood.

Robert Elliott, born 7 March 1887, Richmond, married
Ethel M Hines 25 September 1912, 11 Wellington St, Windsor. Died 24 February 1918, Nhill.

Thomasine Annie, born 24 June 1888, 132 Montague St, Albert Park, married
Harold Beckert, 26 March 1911, Cecil St, South Melbourne.

Albert, born 24 October 1889, 147 Nelson Rd. South Melbourne, married Ethel M Jeffrey 31 August 1912, 27 Duff St. St. Kilda.

John Walter, born 15 November 1890, 128 Montague St. Albert Park, married
Elizabeth Evans 6 May 1913, Church of England, Netherby, Vic.

Herbert Pearce, born 30 April 1892, 13 Banks St, West, South Melbourne, married
Catherine M Freestone, Nhill Methodist Church, 8 January 1916.

Harold Thomas, born 5 June 1893, 142 Ferras St, South Melbourne, married Mary E. Peach 29 November 1913, 13 Lessington St, St. Kilda.

Bessy Melinda, born 19 February 1895, 13 Bank St.W, South Melbourne, married Albert Freestone, 19 February 1922, Brighton, Vic.

Ada, born 18 December 1896, 45 Bridport St, South Melbourne, married Walter C. Freestone, 29 September 1919, North Melbourne.

Arthur Percival, born 27 March 1898, Yarra St, Heidelberg, married Mrs. Jackson.

Figure 6 Thomas and Mary Jane and family

Back: Albert, Thomasine, William, James, John Walter, Robert.

Front: Bessy, Herbert, Ada, Harold, with Mary Jane and Thomas

Figure 37 Mary Jane Dodds with two of her sons (Not Identified)

Figure 38 Thomas Dodds, with child (Not identified)

Thomas went on to raise their large family with Mary Jane. He worked as a boot repairer and ran his own business.

THOMAS DODDS BOOT REPAIR BUSINESS

Figure 39 The Boot Repair Shop

The shop may have been in Bridport St, South Melbourne). We believe that Herbert and Albert are pictured with Thomas.

Their sons Robert Elliot, John Walter and Herbert left Melbourne in 1904, and worked as labourers or 'Teamsters' at Hope Glen, a large holding on the Nhill-Perenna road established by the Dufty family. The Dufty's built a large Mud Brick homestead at 'Hope Glen', complete with a Ballroom, in addition to a large timber and iron multi-stand Shearing Shed, plus many outbuildings. (Garry and Jill Morrison owned the area where the original large house was, and before them Adolph or 'Ardie' Nuske - who purchased it from the Dufty's sometime in the 1950s - and his nephew Don and his wife Faye and family). Ken and Kath have the wooden fireplace surround from Hope Glen in their front room at Grafton.

Robert Elliott married Ethel May Hines at Windsor and still worked in the Lorquon district. He died in February 1918 from kidney failure and was buried in the Nhill cemetery.

Robert Elliott had three children at the time of his death. Hazel 4 years, Thomas Edward 2 years and Robert Harry 1 year. I have not heard of these children even though they would have been cousins to Cliff and Agnes.

Fourth Generation JOHN WALTER DODDS – ELIZABETH EVANS

JOHN WALTER DODDS

Born: 15 November 1890, 128 Montague St., Albert Park.

Married: 6 May 1913, Netherby Church of England.

Died: 1961.

Buried: Nhill Cemetery.

Figure 40 Jack and Elizabeth

ELIZABETH EVANS

Born: 1888

Died: 12 July 1969.

Buried: Nhill Cemetery.

CHILDREN OF JOHN WALTER AND ELIZABETH

Clifford Thomas, born 29 March 1914 at Lorquon West

Agnes Mary, born 4 June 1915.

John Walter (Jack) and Elizabeth Evans married on 6 May 1913 at the Netherby Church of England, having met when they both worked for the Dufty family at Hope Glen. Jack was a teamster and Elizabeth was a maid.

Figure 41 Cliff and Agnes, with Margaret and Harold

They started married life at Pijick but by 1915 had returned to Perenna, buying allotment 74, Parish of Lorquon from Mary Gertrude Dufty. In 1939 the title was discharged to John Walter (Jack) Dodds.

Jack farmed this 1200-acre block, apparently finishing clearing quite a bit of standing Mallee in the 'Back' paddocks during the 1930s and 40s. When his son Cliff left school he worked on the farm with Jack, with their business known as "JW & CT Dodds, Primary Producers, Lorquon West".

Jack and Cliff made the transition from working the land with a team of draught horses to the new development of kerosene-engined tractors in 1944/45, a decision forced on them apparently by the death of several of their horses at the end of the 1944 drought. The horses were sent away on agistment to somewhere in the Western District during the severe drought of 1944 as there was no feed for them at Perenna. When the drought broke, they were returned, but many of them fell ill, apparently due to a too-rapid change of diet back at Perenna.

Jack struggled with the new world of mechanised farming, often as I recall to the angst of Cliff, with many accidents and breakdowns with the new machines. But I do recall his interest in new farming methods, such as visiting the wheat trial plots at the Dahlenburg farm at Salisbury near Nhill and having a clover seed box fitted to a 'Sunder Seeder' (a type of disc seeder) so that they could establish clover medic (Nitrogen Fixing) pastures while they sowed their wheat or oats. Poor rotations which did not include pastures such as clover were one of the many problems that contributed to poor crop yields at the time.

Jack was a committee member of the Nhill A&P Society, which as well as running the Annual Nhill Show. He also used to run the annual 'Nhill A&P Society Crop and Fallow Competition'. I believe that Jack was a judge, or at least used to drive the judges around all the district farms, for these competitions which ran for decades. He and Cliff were proud of the many First, Second and Third prize certificates they themselves won.

Jack, along with a neighbour Eric Berlund, had a strong interest in horse racing, in particular trotting races. For many years there was a trotting event called the 'JW Dodds Handicap' run at the Nhill Show. It was this interest in trotting that brought about his demise as he was killed in a head-on car accident on the Western Highway at Rockbank just out of Melbourne, returning from a day at the trots at Moonee Valley with his good friend and fellow horse lover Bob Lineker of Ballarat. Bob, and his son Lyn, were tailors who used to visit Nhill often to make suits for the men and outfits for the district's women. Jack and Cliff always used to wear a suit when they went out, so the Linekers were an important part of this.

The other great interest away from the farm for Jack, and Cliff, was as supporters of the Collingwood Football Club. They were ardent followers, and apparently would take whatever opportunity they could to get to Melbourne to watch Collingwood play – even riding on the back of a lorry load of wool. This interest can be better understood against the working-class family background in the back streets of Collingwood, Fitzroy and South Melbourne. Collingwood was a team of the working class (Along with Richmond, and South Melbourne)

Figure 42 Three Generations of the Dodds were raised in the house built on the edge of 'Pine Flat" in the late 1920s. It is now derelict.

Jack and Elizabeth lived at Perenna until 1957 when they purchased a house in Nhill at 7 Baker St. They had earlier moved out of the homestead at Perenna into a smaller home in the same farmyard, that was built in the early 1950s. When they moved to Nhill, this house was sold to Roy (Bill) and Beryl Morrison, who moved it a couple of miles east to 'Seven Wells'.

Figure 43 The second house built at 'Pine Flat' for Jack and Elizabeth in the 1950s. Ron and Cat (name unknown!) on the veranda.

Fifth Generation: CLIFFORD THOMAS DODDS and MARGARET IDA GRIMMER

CLIFFORD THOMAS DODDS

Born: 29 March 1914. Nhill, Vic.

Married: 10 April 1943.

Died: 15 September 1988, Melbourne.

Buried: Nhill Cemetery.

MARGARET IDA GRIMMER

Born: 4 November 1917. Wickliffe, Vic.

Died: 19 June 2012, Nhill.

Buried: Nhill cemetery.

CLIFF AND MARGARET'S CHILDREN:

Kenneth John: born 16 January 1944, married Kathleen Mary Standen, 21st May 1966, Clunes, Vic.

Ronald Clifford: born 16 June, 1946, married Mary Janine Smith, 18 April, 1970.

Desmond Harold: born 29 February, 1948, (twins), married Margaret Cristina Costigan.

Shirley Elizabeth: born 29 February 1948 (twins), married Geoffrey Williams.

Graham Jeffrey: born 13 November 1956, married Susan Hilton.

Figure 44 Family Photo - 1950s (Before Graham's arrival)

Figure 45 1970s - Ron, Des, Graham and Shirley with Cliff

Figure 46 1990s. Shirley, Ron, Ken, Graham and Des with Margaret

Sixth Generation: RON DODDS and MARY SMITH

RONALD CLIFFORD DODDS

Born: 16 June 1946

Married: 18 April 1970 in Horsham

MARY JANINE SMITH

Born: 8 September 1948

RON AND MARY'S CHILDREN:

Lisa Jane Dodds: Born 5 December 1972

Russell McGregor Dodds: Born 10 September 1976

Scott Mathew Dodds: Born 8 July 1980

Figure 47 Ron and Mary, Lisa, Scott and Russell (Photo by Brett Wheaton Photography)

Seventh Generation:

Over to you!

Margaret Dodds: Childhood Memories

I think it is most appropriate that Mum's written memoir of her childhood days at Wickliffe should be included here. This piece of her writing was really the inspiration for me to follow on with my story – to shine a light on what might be ordinary day-to-day life in a time and place long gone.

Childhood Memories of Wickliffe in The Western District

BY MARGARET DODDS (NEE GRIMMER) 2005

The saga begins in the early 1920s in Wickliffe, a tiny village in the Western District. Bookaloo was an Aboriginal name my father had given our modest little home which was the third dwelling along one side of a main road. There were about eleven houses in the street and only the Post Office, a quite stately brick building was built on the other side. Halfway along the post office side was the Mechanics Institute Hall, built in memory of soldiers of the 1914-18 war. Beside the hall there stood a monument to the soldiers of that war. I always read my uncle's name on it because he came home with one eye missing and the army would not allow him to have a settler's block, which was provided to returned soldiers who did not have debilitating war injuries. He promptly bought his own block and reared his family on the farm.

The village through my perhaps biased eyes was very picturesque with the whole village encased on flat land sandwiched in between two hills. Against the hill ran the Hopkins River which during my time living there always stopped flowing during summer and was reduced to a series of water holes. All the local kids learned to swim in one of these, taught by a young schoolteacher who taught any of the forty pupils who attended the school who wished to learn to swim. My father Rupert Grimmer had attended the same school when he was young but during his time of fifty-nine years, he never managed to learn to swim. I doubt if many of his generation could swim as it wasn't fashionable in that era. Occasionally in wet winters the banks overflowed, and the water covered the flat and even threatened to come in the houses. It was only by strategic use of sandbags that it was kept back from the dwellings.

The village probably was sited there because there were two large lagoons across the road from the hotel. The only known transport in early days was by horse and buggy and the lagoons were a source of good drinking water for stock. Those living in the street each had their own cows to milk and the cows were allowed to run on the common as it was called.

The great depression settled in the late twenties and early thirties and money was very scarce. Scores of men were swagmen walking the country looking for work. Dad took on one to help him with his 15 odd Clydesdale horses and growing wheat. He stayed with us for some time sleeping in the shed and helping with farm work.

Mum had five children and two men to cook for.

The lagoons just across from the hotel were full of little redfin perch. One summer when they were drying up Dad took a team of about eight big draught horses and drove them in the water till it was all muddied and the fish had to come up for air. We scooped them off in buckets and had the two wash troughs full of red perch. There was no such thing as refrigeration in those days. So after keeping enough for us the surplus fish were given away.

At the side of our house over the road was a vacant block. It had not been fenced and it was called the "green."

Late in the evening when children's jobs finished several of them from the street houses came there to play. There was always something going on. Everybody played football, cricket, brandy (with a tennis ball), catchy and sailing down the hill in cars. The fun broke up when somebody's mother called the children in for tea or bed or whatever. We were all in bed by nine o'clock at nights. Lights were always out by then. The lights were powered by kerosene. Each evening the lamp had to be filled with kerosene and the wick trimmed. When the wick was burnt out you bought a new one at the general store in the centre of the village.

All the things we think are normal were missing. No cars, aeroplanes, refrigerators, television, freezers and electricity; very few had bathrooms with running water, no doonas, no washing powder, no plastic goods, or detergent, only soap – often homemade, no toilet soap just plain old bars of 'velvet' or 'preservene 'soap – no washing machines. Mum was up at 5 a.m. to get the stove, which was wood powered, alight. Every day when I came home from school, I had to get enough kindling wood to light the stove for the next day. It had to be dry enough to burn. Usually, it was a bundle of gum twigs tied together. You started the fire with a match to crumpled paper and the dry twigs readily caught fire with a lovely crackle. Then you had lightly split other wood to go on from there.

The men had to get up early to feed and harness the working horses. Each horse had to have collar 'hames' and 'winkers' put on him. While he was eating the "bit" that went through his mouth to guide the reins with was put in place later.

It was an endless job for farmers - cutting chaff. Horses work better if they are fed well. Sometimes, the men walked behind the horse to the paddock driving them with the reins. They were hitched to a plough or another implement when they arrived where the crop would be grown. Most of my father's cropping was grown on rented land. My father rented land from the "squatter" who owned 40 thousand acres of land. Horses and men worked from daylight to dark ploughing, harrowing, seeding and harvesting.

One of the joys of childhood for me was standing on the board of the combine which seeded the crop and watching the dozens of birds that swooped in as soon as the earth was tilled so as to pick up all the tasty fat worms which the sods of earth uncovered. I see in my mind's eye, the plovers, the magpies and all the little birds that followed closely on behind the machines. Curlews came and went. One great day I remember seeing four native companions (Brolga) doing a very elegant and clever dance. They were very graceful creatures and at that time I couldn't ally myself with father's thinking that he had to shoot them; at least one to scare the others off as they ate the grain. The idea was that if it was tied spread out to the fence it would scare the others away from the seed. Now I see it the other way, of how little he earned for all his hard work when the crop returned six bags to the acre. One day we caught a little bandicoot and kept him in a box with straw for a few weeks feeding him scraps of bacon until we sent him to the Healesville Sanctuary to be looked after by a naturalist, Dr David Fleay.

There were lots of old redgum and she-oak logs lying around the paddock and as children we were allowed to make fires out of them. One year when mushrooms were plentiful on the untilled ground, we collected lots of them and cooked them on the coals of a log fire. Mum had a huge old black cast iron pan with a handle like a basket and to use modern language - they were yummy for lunch.

Sunday was a day of rest. We five were always sent to church – Presbyterian – a Scottish based religion. A former Scotsman squatter who took up the forty-thousand-acre station property built and endowed the beautiful little old blue-stone church in the valley. He could climb the tower of his Gothic built homestead and look over the two-mile distance between his home and church. Mum always stayed home to cook the Sunday roast dinner. You could never work in the garden on Sunday or knit or use the sewing machine. You attended church at 11 a.m. My two brothers at one stage rang the bells – the first bell at 10.45 am and the second bell at 11 am - while the minister walked up the aisle of the church to the pulpit. The bell was housed in a lovely little blue stone tower. The whole of the church was blue stone, and the furnishings were all well carved from beautiful wood. Sometimes, I went the back way to Sunday school. There was a horrible little old dam on the fence line with water in it. Dad told me that in that dam there lived a Jenny Green tooth who when you went close to the water would jump out and grab you. For a long time, I walked in a half circle away from the dam. It had a good effect as there was no danger of me being drowned in the dam.

Quite recently my younger brother who is now Eighty-Five years told me he read about a' Jenny Green Tooth 'being housed in a waterway in the north of England so that could have been where it originated.

A beautiful little kindergarten room had been arranged in the back section of the Manse which housed the minister, his wife and young family. There were about fifteen chairs, blonde ones in a ring. They were for the tots to sit, a large chair for the young teacher who would have been around twenty years old. We loved her as she told us stories. She still lives in the area with lots of descendants who have dispersed near and far. A year or two ago she reached the 100-year milestone and to my knowledge she still enjoys her life. I didn't mention there was a lovely red velvet rocking chair for each child to sit in on the Sunday of their birthday week while we all sang happy birthday. Also, in a corner was quite a large wooden tray filled with saw dust with cardboard cut outs of Arabs, animals, camels mainly, and biblical characters where little folk could sit around and build up in pretence, the essence of the story we had just been told.

Once a year there was the Sunday school anniversary when all the parents came to church. The minister presented each child with a book and if they were one of the lucky ones had a paper stuck inside the cover saying 1st, 2nd 3rd prize.

The books we had at home were only printed in black and white like the daily papers used to be until a couple of years ago. These books had coloured covers and interesting pretty pictures. The superintendent was a mature thin lady. She never missed a Sunday for years and always she was dressed in black from head to toe. She had a flat looking hat sitting atop a big bun of hair at the back above her neck. The dress was complemented by a large white collar of lace and it went all around her neck. It was reinforced with bits of whale bone to hold it against her neck.

When collection was taken up in Sunday school they sang "Hear the pennies dropping – listen when they fall." Then they marched each to their own rooms, the sermon began, and church carried on with adults only.

The minister only resided in the village a few years and then moved on. The elders of the church chose the next minister by what was called "preaching for a call." The heads of the church in Melbourne sent out young, trained ministers once a week. They preached at church, met the heads after church who took him to Sunday lunch and after a few weeks they decided who would get the "call." Usually, the minister fitted in quite well with the folk of the village.

The squatter who built the church endowed it with enough money to pay the salary of the minister but a few years ago when there were only eight people attending church, it was reluctantly decided to close it. But they never did. They hold church now with a congregation of five in the valley – a potent reminder to those who spent their childhood there. The pews all sit there very carefully preserved along with the organ, ornate pulpit and honour roll from World War 1914-18.

If it was a sunny Sunday, my sister and I were sent with flowers to take to the cemetery and place on Grandma's grave. She had passed on in 1927 at the age of 63. She had reared a family of eight children, losing one. Around two or three of her sons and one daughter had settled nearby when they married. Eventually, there were 19 cousins living in the area.

I never met my grandfather. He had passed on before my era. Grandma lived with two single daughters. The youngest one was sewing mistress at the local school and the other one was a trained nurse engaged to a station master. When Grandma fell ill, they found it necessary for the trained nurse to come home and care for her. So ended her plans of married life. The younger aunt lost her fiancée in the Great War. They lived on well into their eighties with church as one of their main interests in life. They also loved flowers and gardens.

They lived on 120 acres of land which was virgin country taken up by their grandfather. That land is still in the hands of my brother and his sons after four generations.

One of the biggest trips we did was with two horses pulling the buggy. We travelled to the home of my mother's aunt and uncle. They only lived 10 miles away. We made the trip about three or four times a year arriving just before lunch and leaving after tea. They were also farmers. We got our first telephone installed in our home in December 1926. I know this because Dad wrote it on the wall beside it. My older brother rode a small pony, Nigger9 along beside us. The buggy had 2 upholstered seats – Dad, Mum and the baby in the front – the two girls and one boy sat on the back seat. I mention the telephone because then we could ring her up and tell her when we were coming. They had a lovely vegetable and flower garden. There was a light on either side of the buggy for night travel.

The one yearly special event was held in October – the Flower Show. The hall was filled with flowers for judging. There were also prizes for house gardens judged by a nurseryman from a local large town 40 miles away. Sometimes we got a prize for our garden. I remember the burst of energy required at the last minute to get it ship-shape with no visible weeds. When we grew bigger my sister and I were secretaries of the Flower Show. There were sections for different grain such as barley, wheat and oats.

Dad got a master plan that as we lived at the base of a substantial hill, he could come one-third of the way down the hill – and scoop out a dam using horses with a scoop. He diligently worked on this scooping out of a lot of sticky yellow clay. He heaped it all on the lower side of the dam making quite an imposing embankment. When we got a decent rain the dam soon filled but alas as soon as the water got against the yellow clay bank it oozed straight through and went on down the hill in yellow streams. There were children there for days with no boots in this sticky, oozy bank trying to stop the flow but to no avail. The big idea was to pipe the water down to the vegetable beds. Eventually we wrote if off. Since I grew up, I never looked to see if the old dam was still there.

---

9 This word is left unedited – its usage at the time was not seem as unacceptable, as it is today.

There were always pets around the place. My very own first pet was a white Corella cockatoo. When they were leaving the lady at the hotel gave me this fully grown cocky – snow white – talkative, cunning and mischievous. He learned to hold his head on one side and lift up the little trap door and carefully edge himself out. If he wasn't discovered by anybody, he could whip out a whole row of strawberries and lay the bushes on their sides. We soon ended the high jinks with a brick behind the little door.

Cocky's next move was to bite the wire on his cage and when he got enough broken through, he could ease himself out. His cage soon looked decrepit with bits of board nailed across the front. Dad got the idea of inventing a stand with a roost on top, put a chain on his leg and so tie him up. For two or three days he solemnly sat on the stand then one bright morning he hit on the idea that he could undo the screw that let his chain fall off his leg. After that the stand had to be discarded as the band would be undone before you could go back inside. At night a canvass was hung over his cage to let him know it was bedtime. If you bumped into his cage in the dark, he would yell out "Go to Hell" or "Have a swim." The language he had learned in the hotel.

When I returned home from first term holidays at school, (Alexandra College, Hamilton) Cocky was no longer there. Dad said he gave him to a fellow in the next village.

As far back as I can remember we always had lovely little ponies to ride. Mum, Dad and the baby drove to Warrnambool for a holiday. When they came back, they told the four of us there was a present for us tied to a fence beyond the last house. We raced there and there was this lovely little pony. She was a bay colour and very fat. She had a bridle and across the nose band was printed "PLAYTOY." She had been broken in to ride and the four of us had hours of pleasant memories of riding her and carrying messages. One day I recall when my sister and I rode PLAYTOY seven miles to put a cheque in the bank.

Later on when I was bigger, we used to go "coursing" which was chasing rabbits with a greyhound as rabbits had reached pestilent numbers. My older brother set rabbit traps. When he went around them in the dark, I could go with him to carry a kerosene lantern which provided a reasonable light. Rabbits were known as underground mutton. When we came home from school for lunch the menu was often extended with rabbit in white sauce or baked.

Billy and Willy were two motherless lambs I bottle-fed from early life. Eventually they grew to sheep and were put in with the mob. Dad had a little plantation of young sugar gums up the hill; I used to tie the two big lambs up to a tree on the way to school so they could nibble any grass around. One day I was kept in after school for talking in school and the teacher told me to write out 200 words for punishment. I meekly asked if I could untangle the lambs before I set to write them. His reply was "Get out of my sight."

The same teacher had put his new pocketknife on the front of his desk and as he couldn't see it when he was about to dismiss the 40 odd children, he gave the ultimatum that no one was going to leave the school until the one who took the knife owned up. "Does anybody know who took it?" he asked. After a time, a lanky Japanese boy put up his hand and said, "Please sir, I know someone who knows who took it?"

"Well Jack who knows who took it?" asked the teacher.

"Please sir, God knows!"

Years after, someone found the rusted knife in a plot of dirt near the chimney.

The Horses

The draught horses were lovely animals. They had a big stable at the end of the house block. When they came home at night, they each went to their own stall. Tommy was a lovely old black horse with a white blaze on his face. They had measured rations to each feeder after working all day. They had a kerosene bucket of chaff, a six-pound jam tin of oats and a tin of bran each. Sometimes in summer each horse was clipped. They were sometimes clipped halfway up their side or else all over. It took two men to do this. One held the clippers while the other twined the handle of the machine. During hot weather hard working horses sweated a lot. It looked like soap lather when they broke into a sweat.

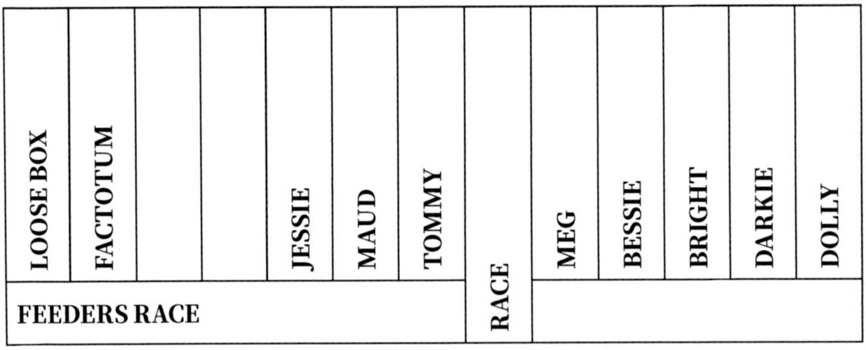

It didn't seem a week went by without chaff having to be cut. The chaff from the chaff cutter was elevated into the chaff house and if it was from hay that was a bit green, it got hot. It was possible to cause a fire, but it never did. Occasionally, there was a new little foal which always looked so very dainty.

Tommy was the reliable old cart horse. If he lost his shoe and he was let out he would go to the blacksmith's shop and wait till they took him in and got another.

On one New Year's Eve, the local lads played a bit of mischief by whitewashing old Tommy all over. He looked a sorry sight on New Year's Day when he stood out in the horse group with some bits of hair black and other bits white.

Jerry was the old 'Gig' horse. He was bay coloured, that is, a deep reddish brown. His legs were very long, and he had black hair around his feet. Sometimes, when we drove him, Dad would shoot a rabbit and when the gun made the bang, Jerry would always jump off the ground. Sometimes after he was taken out of the gig and unharnessed, he rolled in the dust on his back – his way of getting the sweat off his back.

When my sister, six years older than I, left school, she used to drive Jerry in the gig with the hot lunch Mum had cooked to the men working in the paddock.

One day my brother and I were driving to the next village – Lake Bolac, ten miles away. Rocket, an awkward looking horse was yoked to the gig. Charlie, my older brother had just broken him in, to harness. After we had travelled five miles, he lay on his side on the road and refused to get up in spite of Charlie's taunts. However, the next urging caused activity again. Men and boys wore felt hats in those days. Charlie put some water in his hat from a roadside puddle. He gently poured it down one ear that was sticking up and Rocket was on his legs in a jiffy – all ready to pull the gig again.

Mum's written story of memories from her childhood ends here.

About the Author: Ron Dodds

Ron Dodds was born at Nhill, Victoria in 1946, the second son of Cliff and Margaret Dodds. He, and his four siblings were raised on their family farm at Perenna, thirty-two kilometres north of Nhill.

He was educated at the small single teacher Perenna State School, which was adjacent to their farm, then to High School at Nhill, completing his matriculation in 1963.

Upon leaving school he joined his elder brother Ken in 1964 on the family farm, thus commencing a career in agriculture that lasted until 2000. Ron was joined on the farm by his wife Mary upon their marriage in 1970.

While not progressing to tertiary education, Ron had a strong interest in scientific research and its application to the fast-changing farming scene at the time. The family partnership were early adopters of many new practices in agriculture, and his interest led him to several bodies external to the farm which were dedicated to the adoption of new technology. This included The Wheat Research Foundation of Victoria (Chair), The Victorian Institute for Dryland Agriculture Advisory Committee (Chair), the Wimmera Conservation Farming Association, and the West Wimmera Farm Tree Group.

Ron's interests did extend beyond this technical realm – to just being part of his community and all that involved. Particularly local sport, but also to enjoying living in a rural community and the lifestyle that afforded. Like he and his siblings before him, Ron and Mary's three children Lisa, Russell and Scott were all raised in that rural setting.

Rons interests external to the farm, and his interest in nature, did lead him in a different direction, joining the non-government conservation organisation Greening Australia in 1993. This eventually led to his leaving his place of birth at Perenna after fifty-four years. Ron commenced this work as a part time facilitator in the Wimmera, but extended to a regional management role, and then statewide, culminating in the role of Victorian State Manager, the position he held at the time of his retirement in 2013.

ISBN: 978-0-646-71096-9

Layout Design by: Wimmera Design and Print, Horsham, Victoria, AUS

Date of Publication: July 2025

Every effort has been made to ensure the accuracy of names, dates, and details contained in this publication. The information presented is based on available records, research, and personal recollections. Despite this, there may be occasional inaccuracies or omissions due to the limitations of historical documentation.

Made in United States
Orlando, FL
05 November 2026

90231961R00119